U·X·L
Encyclopedia
of Science

FARMINGDALE PUBLIC LIBRARY
116 MERRITTS ROAD
FARMINGDALE, N.Y. 11735

U·X·L Encyclopedia of Science

Second Edition

Volume 3: Cat-Cy

Rob Nagel, Editor

GALE GROUP
THOMSON LEARNING

Detroit • New York • San Diego • San Francisco
Boston • New Haven, Conn. • Waterville, Maine
London • Munich

U·X·L Encyclopedia of Science
Second Edition

Rob Nagel, *Editor*

Staff

Elizabeth Shaw Grunow, *U•X•L Editor*

Julie Carnagie, *Contributing Editor*

Carol DeKane Nagel, *U•X•L Managing Editor*

Thomas L. Romig, *U•X•L Publisher*

Shalice Shah-Caldwell, *Permissions Associate (Pictures)*

Robyn Young, *Imaging and Multimedia Content Editor*

Rita Wimberley, *Senior Buyer*

Pamela A. E. Galbreath, *Senior Art Design*er

Michelle Cadorée, *Indexing*

GGS Information Services, *Typesetting*

On the front cover: Nikola Tesla with one of his generators, reproduced by permission of the Granger Collection.

On the back cover: The flow of red blood cells through blood vessels, reproduced by permission of Phototake.

Library of Congress Cataloging-in-Publication Data

U-X-L encyclopedia of science.—2nd ed. / Rob Nagel, editor
p.cm.
Includes bibliographical references and indexes.
Contents: v.1. A-As — v.2. At-Car — v.3. Cat-Cy — v.4. D-Em — v.5. En-G — v.6. H-Mar — v.7. Mas-O — v.8. P-Ra — v.9. Re-St — v.10. Su-Z.
Summary: Includes 600 topics in the life, earth, and physical sciences as well as in engineering, technology, math, environmental science, and psychology.
ISBN 0-7876-5432-9 (set : acid-free paper) — ISBN 0-7876-5433-7 (v.1 : acid-free paper) — ISBN 0-7876-5434-5 (v.2 : acid-free paper) — ISBN 0-7876-5435-3 (v.3 : acid-free paper) — ISBN 0-7876-5436-1 (v.4 : acid-free paper) — ISBN 0-7876-5437-X (v.5 : acid-free paper) — ISBN 0-7876-5438-8 (v.6 : acid-free paper) — ISBN 0-7876-5439-6 (v.7 : acid-free paper) — ISBN 0-7876-5440-X (v.8 : acid-free paper) — ISBN 0-7876-5441-8 (v.9 : acid-free paper) — ISBN 0-7876-5775-1 (v.10 : acid-free paper)

1. Science-Encyclopedias, Juvenile. 2. Technology-Encyclopedias, Juvenile. [1. Science-Encyclopedias. 2. Technology-Encyclopedias.] I. Title: UXL encyclopedia of science. II. Nagel, Rob.
Q121.U18 2001
503-dc21

2001035562

This publication is a creative work fully protected by all applicable copyright laws, as well as by misappropriation, trade secret, unfair competition, and other applicable laws. The editors of this work have added value to the underlying factual material herein through one or more of the following: unique and original selection, coordination, expression, arrangement, and classification of the information. All rights to this publication will be vigorously defended.

Copyright © 2002   U•X•L, an imprint of The Gale Group

All rights reserved, including the right of reproduction in whole or in part in any form.

Printed in the United States of America

10 9 8 7 6 5 4 3 2 1

Table of Contents

Reader's Guide vii

Entries by Scientific Field ix

Volume 1: A-As 1
    Where to Learn More xxxi
    Index xxxv

Volume 2: At-Car 211
    Where to Learn More xxxi
    Index xxxv

Volume 3: Cat-Cy 413
    Where to Learn More xxxi
    Index xxxv

Volume 4: D-Em 611
    Where to Learn More xxxi
    Index xxxv

Volume 5: En-G 793
    Where to Learn More xxxi
    Index xxxv

Volume 6: H-Mar 1027
    Where to Learn More xxxi
    Index xxxv

Volume 7: Mas-O 1235
    Where to Learn More xxxi
    Index xxxv

U·X·L Encyclopedia of Science, 2nd Edition

Contents

Volume 8: P-Ra 1457
    Where to Learn More . . . . **xxxi**
    Index . . . . **xxxv**

Volume 9: Re-St 1647
    Where to Learn More . . . . **xxxi**
    Index . . . . **xxxv**

Volume 10: Su-Z 1829
    Where to Learn More . . . . **xxxi**
    Index . . . . **xxxv**

Reader's Guide

Demystify scientific theories, controversies, discoveries, and phenomena with the *U•X•L Encyclopedia of Science,* Second Edition.

This alphabetically organized ten-volume set opens up the entire world of science in clear, nontechnical language. More than 600 entries—an increase of more than 10 percent from the first edition—provide fascinating facts covering the entire spectrum of science. This second edition features more than 50 new entries and more than 100 updated entries. These informative essays range from 250 to 2,500 words, many of which include helpful sidebar boxes that highlight fascinating facts and phenomena. Topics profiled are related to the physical, life, and earth sciences, as well as to math, psychology, engineering, technology, and the environment.

In addition to solid information, the *Encyclopedia* also provides these features:

- "Words to Know" boxes that define commonly used terms
- Extensive cross references that lead directly to related entries
- A table of contents by scientific field that organizes the entries
- More than 600 color and black-and-white photos and technical drawings
- Sources for further study, including books, magazines, and Web sites

Each volume concludes with a cumulative subject index, making it easy to locate quickly the theories, people, objects, and inventions discussed throughout the *U•X•L Encyclopedia of Science,* Second Edition.

Reader's Guide

Suggestions

We welcome any comments on this work and suggestions for entries to feature in future editions of *U•X•L Encyclopedia of Science*. Please write: Editors, *U•X•L Encyclopedia of Science,* U•X•L, Gale Group, 27500 Drake Road, Farmington Hills, Michigan, 48331-3535; call toll-free: 800-877-4253; fax to: 248-699-8097; or send an e-mail via www.galegroup.com.

Entries by Scientific Field

Boldface indicates volume numbers.

Acoustics

Acoustics	**1**:17
Compact disc	**3**:531
Diffraction	**4**:648
Echolocation	**4**:720
Magnetic recording/ audiocassette	**6**:1209
Sonar	**9**:1770
Ultrasonics	**10**:1941
Video recording	**10**:1968

Aerodynamics

Aerodynamics	**1**:39
Fluid dynamics	**5**:882

Aeronautical engineering

Aircraft	**1**:74
Atmosphere observation	**2**:215
Balloon	**1**:261
Jet engine	**6**:1143
Rockets and missiles	**9**:1693

Aerospace engineering

International Ultraviolet Explorer	**6**:1120
Rockets and missiles	**9**:1693
Satellite	**9**:1707
Spacecraft, manned	**9**:1777
Space probe	**9**:1783
Space station, international	**9**:1788
Telescope	**10**:1869

Agriculture

Agriculture	**1**:62
Agrochemical	**1**:65
Aquaculture	**1**:166
Biotechnology	**2**:309
Cotton	**3**:577
Crops	**3**:582
DDT (dichlorodiphenyl-trichloroethane)	**4**:619
Drift net	**4**:680
Forestry	**5**:901
Genetic engineering	**5**:973
Organic farming	**7**:1431
Slash-and-burn agriculture	**9**:1743
Soil	**9**:1758

Anatomy and physiology

Anatomy	**1**:138
Blood	**2**:326

Entries by Scientific Field

Brain	2:337
Cholesterol	3:469
Chromosome	3:472
Circulatory system	3:480
Digestive system	4:653
Ear	4:693
Endocrine system	5:796
Excretory system	5:839
Eye	5:848
Heart	6:1037
Human Genome Project	6:1060
Immune system	6:1082
Integumentary system	6:1109
Lymphatic system	6:1198
Muscular system	7:1309
Nervous system	7:1333
Physiology	8:1516
Reproductive system	9:1667
Respiratory system	9:1677
Skeletal system	9:1739
Smell	9:1750
Speech	9:1796
Taste	10:1861
Touch	10:1903

Anesthesiology

Alternative medicine	1:118
Anesthesia	1:142

Animal husbandry

Agrochemical	1:65
Biotechnology	2:309
Crops	3:582
Genetic engineering	5:973
Organic farming	7:1431

Anthropology

Archaeoastronomy	1:171
Dating techniques	4:616
Forensic science	5:898
Gerontology	5:999
Human evolution	6:1054
Mounds, earthen	7:1298
Petroglyphs and pictographs	8:1491

Aquaculture

Aquaculture	1:166
Crops	3:582
Drift net	4:680
Fish	5:875

Archaeology

Archaeoastronomy	1:171
Archaeology	1:173
Dating techniques	4:616
Fossil and fossilization	5:917
Half-life	6:1027
Nautical archaeology	7:1323
Petroglyphs and pictographs	8:1491

Artificial intelligence

Artificial intelligence	1:188
Automation	2:242

Astronomy

Archaeoastronomy	1:171
Asteroid	1:200
Astrophysics	1:207
Big bang theory	2:273
Binary star	2:276
Black hole	2:322
Brown dwarf	2:358
Calendar	2:372
Celestial mechanics	3:423
Comet	3:527
Constellation	3:558
Cosmic ray	3:571

Cosmology	3:574
Dark matter	4:613
Earth (planet)	4:698
Eclipse	4:723
Extrasolar planet	5:847
Galaxy	5:941
Gamma ray	5:949
Gamma-ray burst	5:952
Gravity and gravitation	5:1012
Infrared astronomy	6:1100
International Ultraviolet Explorer	6:1120
Interstellar matter	6:1130
Jupiter (planet)	6:1146
Light-year	6:1190
Mars (planet)	6:1228
Mercury (planet)	7:1250
Meteor and meteorite	7:1262
Moon	7:1294
Nebula	7:1327
Neptune (planet)	7:1330
Neutron star	7:1339
Nova	7:1359
Orbit	7:1426
Pluto (planet)	8:1539
Quasar	8:1609
Radio astronomy	8:1633
Red giant	9:1653
Redshift	9:1654
Satellite	9:1707
Saturn (planet)	9:1708
Seasons	9:1726
Solar system	9:1762
Space	9:1776
Spacecraft, manned	9:1777
Space probe	9:1783
Space station, international	9:1788
Star	9:1801
Starburst galaxy	9:1806
Star cluster	9:1808
Stellar magnetic fields	9:1820
Sun	10:1844
Supernova	10:1852
Telescope	10:1869
Ultraviolet astronomy	10:1943
Uranus (planet)	10:1952
Variable star	10:1963
Venus (planet)	10:1964
White dwarf	10:2027
X-ray astronomy	10:2038

Astrophysics

Astrophysics	1:207
Big bang theory	2:273
Binary star	2:276
Black hole	2:322
Brown dwarf	2:358
Celestial mechanics	3:423
Cosmic ray	3:571
Cosmology	3:574
Dark matter	4:613
Galaxy	5:941
Gamma ray	5:949
Gamma-ray burst	5:952
Gravity and gravitation	5:1012
Infrared astronomy	6:1100
International Ultraviolet Explorer	6:1120
Interstellar matter	6:1130
Light-year	6:1190
Neutron star	7:1339
Orbit	7:1426
Quasar	8:1609
Radio astronomy	8:1633
Red giant	9:1653
Redshift	9:1654
Space	9:1776
Star	9:1801
Starburst galaxy	9:1806
Star cluster	9:1808
Stellar magnetic fields	9:1820
Sun	10:1844

Entries by Scientific Field

Entries by Scientific Field

Supernova **10**:1852
Ultraviolet astronomy **10**:1943
Uranus (planet) **10**:1952
Variable star **10**:1963
White dwarf **10**:2027
X-ray astronomy **10**:2038

Atomic/Nuclear physics

Actinides **1**:23
Alkali metals **1**:99
Alkali earth metals **1**:102
Alternative energy sources **1**:111
Antiparticle **1**:163
Atom **2**:226
Atomic mass **2**:229
Atomic theory **2**:232
Chemical bond **3**:453
Dating techniques **4**:616
Electron **4**:768
Half-life **6**:1027
Ionization **6**:1135
Isotope **6**:1141
Lanthanides **6**:1163
Mole (measurement) **7**:1282
Molecule **7**:1285
Neutron **7**:1337
Noble gases **7**:1349
Nuclear fission **7**:1361
Nuclear fusion **7**:1366
Nuclear medicine **7**:1372
Nuclear power **7**:1374
Nuclear weapons **7**:1381
Particle accelerators **8**:1475
Quantum mechanics **8**:1607
Radiation **8**:1619
Radiation exposure **8**:1621
Radiology **8**:1637
Subatomic particles **10**:1829
X ray **10**:2033

Automotive engineering

Automobile **2**:245
Diesel engine **4**:646
Internal-combustion engine **6**:1117

Bacteriology

Bacteria **2**:253
Biological warfare **2**:287
Disease **4**:669
Legionnaire's disease **6**:1179

Ballistics

Ballistics **2**:260
Nuclear weapons **7**:1381
Rockets and missiles **9**:1693

Biochemistry

Amino acid **1**:130
Biochemistry **2**:279
Carbohydrate **2**:387
Cell **3**:428
Cholesterol **3**:469
Enzyme **5**:812
Fermentation **5**:864
Hormones **6**:1050
Human Genome Project **6**:1060
Lipids **6**:1191
Metabolism **7**:1255
Nucleic acid **7**:1387
Osmosis **7**:1436
Photosynthesis **8**:1505
Proteins **8**:1586
Respiration **9**:1672
Vitamin **10**:1981
Yeast **10**:2043

Entries by Scientific Field

Biology

Adaptation	1:26
Algae	1:91
Amino acid	1:130
Amoeba	1:131
Amphibians	1:134
Anatomy	1:138
Animal	1:145
Antibody and antigen	1:159
Arachnids	1:168
Arthropods	1:183
Bacteria	2:253
Behavior	2:270
Biochemistry	2:279
Biodegradable	2:280
Biodiversity	2:281
Biological warfare	2:287
Biology	2:290
Biome	2:293
Biophysics	2:302
Biosphere	2:304
Biotechnology	2:309
Birds	2:312
Birth	2:315
Birth defects	2:319
Blood	2:326
Botany	2:334
Brain	2:337
Butterflies	2:364
Canines	2:382
Carbohydrate	2:387
Carcinogen	2:406
Cell	3:428
Cellulose	3:442
Cetaceans	3:448
Cholesterol	3:469
Chromosome	3:472
Circulatory system	3:480
Clone and cloning	3:484
Cockroaches	3:505
Coelacanth	3:508
Contraception	3:562
Coral	3:566
Crustaceans	3:590
Cryobiology	3:593
Digestive system	4:653
Dinosaur	4:658
Disease	4:669
Ear	4:693
Embryo and embryonic development	4:785
Endocrine system	5:796
Enzyme	5:812
Eutrophication	5:828
Evolution	5:832
Excretory system	5:839
Eye	5:848
Felines	5:855
Fermentation	5:864
Fertilization	5:867
Fish	5:875
Flower	5:878
Forestry	5:901
Forests	5:907
Fungi	5:930
Genetic disorders	5:966
Genetic engineering	5:973
Genetics	5:980
Heart	6:1037
Hibernation	6:1046
Hormones	6:1050
Horticulture	6:1053
Human Genome Project	6:1060
Human evolution	6:1054
Immune system	6:1082
Indicator species	6:1090
Insects	6:1103
Integumentary system	6:1109
Invertebrates	6:1133
Kangaroos and wallabies	6:1153
Leaf	6:1172
Lipids	6:1191
Lymphatic system	6:1198
Mammals	6:1222

Entries by Scientific Field

Mendelian laws of inheritance	**7:**1246	Vaccine	**10:**1957
Metabolism	**7:**1255	Vertebrates	**10:**1967
Metamorphosis	**7:**1259	Virus	**10:**1974
Migration (animals)	**7:**1271	Vitamin	**10:**1981
Molecular biology	**7:**1283	Wetlands	**10:**2024
Mollusks	**7:**1288	Yeast	**10:**2043
Muscular system	**7:**1309		
Mutation	**7:**1314		
Nervous system	**7:**1333		

Biomedical engineering

Electrocardiogram	**4:**751
Radiology	**8:**1637

Nucleic acid	**7:**1387		
Osmosis	**7:**1436		
Parasites	**8:**1467		
Photosynthesis	**8:**1505		
Phototropism	**8:**1508		

Biotechnology

Biotechnology	**2:**309
Brewing	**2:**352
Fermentation	**5:**864
Vaccine	**10:**1957

Physiology	**8:**1516		
Plague	**8:**1518		
Plankton	**8:**1520		
Plant	**8:**1522		
Primates	**8:**1571		
Proteins	**8:**1586		

Botany

Botany	**2:**334
Cellulose	**3:**442
Cocaine	**3:**501
Cotton	**3:**577
Flower	**5:**878
Forestry	**5:**901
Forests	**5:**907
Horticulture	**6:**1053
Leaf	**6:**1172
Marijuana	**6:**1224
Photosynthesis	**8:**1505
Phototropism	**8:**1508
Plant	**8:**1522
Seed	**9:**1729
Tree	**10:**1927

Protozoa	**8:**1590		
Puberty	**8:**1599		
Rain forest	**8:**1641		
Reproduction	**9:**1664		
Reproductive system	**9:**1667		
Reptiles	**9:**1670		
Respiration	**9:**1672		
Respiratory system	**9:**1677		
Rh factor	**9:**1683		
Seed	**9:**1729		
Sexually transmitted diseases	**9:**1735		
Skeletal system	**9:**1739		
Smell	**9:**1750		
Snakes	**9:**1752		
Speech	**9:**1796		
Sponges	**9:**1799		
Taste	**10:**1861		
Touch	**10:**1903		

Cartography

Cartography	**2:**410
Geologic map	**5:**986

Tree	**10:**1927
Tumor	**10:**1934

Entries by Scientific Field

Cellular biology

Amino acid	**1**:130
Carbohydrate	**2**:387
Cell	**3**:428
Cholesterol	**3**:469
Chromosome	**3**:472
Genetics	**5**:980
Lipids	**6**:1191
Osmosis	**7**:1436
Proteins	**8**:1586

Chemistry

Acids and bases	**1**:14
Actinides	**1**:23
Aerosols	**1**:43
Agent Orange	**1**:54
Agrochemical	**1**:65
Alchemy	**1**:82
Alcohols	**1**:88
Alkali metals	**1**:99
Alkaline earth metals	**1**:102
Aluminum family	**1**:122
Atom	**2**:226
Atomic mass	**2**:229
Atomic theory	**2**:232
Biochemistry	**2**:279
Carbon dioxide	**2**:393
Carbon family	**2**:395
Carbon monoxide	**2**:403
Catalyst and catalysis	**2**:413
Chemical bond	**3**:453
Chemical w\arfare	**3**:457
Chemistry	**3**:463
Colloid	**3**:515
Combustion	**3**:522
Composite materials	**3**:536
Compound, chemical	**3**:541
Crystal	**3**:601
Cyclamate	**3**:608
DDT (dichlorodiphenyl-trichloroethane)	**4**:619
Diffusion	**4**:651
Dioxin	**4**:667
Distillation	**4**:675
Dyes and pigments	**4**:686
Electrolysis	**4**:755
Element, chemical	**4**:774
Enzyme	**5**:812
Equation, chemical	**5**:815
Equilibrium, chemical	**5**:817
Explosives	**5**:843
Fermentation	**5**:864
Filtration	**5**:872
Formula, chemical	**5**:914
Halogens	**6**:1030
Hormones	**6**:1050
Hydrogen	**6**:1068
Industrial minerals	**6**:1092
Ionization	**6**:1135
Isotope	**6**:1141
Lanthanides	**6**:1163
Lipids	**6**:1191
Metabolism	**7**:1255
Mole (measurement)	**7**:1282
Molecule	**7**:1285
Nitrogen family	**7**:1344
Noble gases	**7**:1349
Nucleic acid	**7**:1387
Osmosis	**7**:1436
Oxidation-reduction reaction	**7**:1439
Oxygen family	**7**:1442
Ozone	**7**:1450
Periodic table	**8**:1486
pH	**8**:1495
Photochemistry	**8**:1498
Photosynthesis	**8**:1505
Plastics	**8**:1532
Poisons and toxins	**8**:1542
Polymer	**8**:1563
Proteins	**8**:1586
Qualitative analysis	**8**:1603
Quantitative analysis	**8**:1604

Entries by Scientific Field

Reaction, chemical	9:1647
Respiration	9:1672
Soaps and detergents	9:1756
Solution	9:1767
Transition elements	10:1913
Vitamin	10:1981
Yeast	10:2043

Civil engineering

Bridges	2:354
Canal	2:376
Dam	4:611
Lock	6:1192

Climatology

Global climate	5:1006
Ice ages	6:1075
Seasons	9:1726

Communications/ Graphic arts

Antenna	1:153
CAD/CAM	2:369
Cellular/digital technology	3:439
Compact disc	3:531
Computer software	3:549
DVD technology	4:684
Hologram and holography	6:1048
Internet	6:1123
Magnetic recording/ audiocassette	6:1209
Microwave communication	7:1268
Petroglyphs and pictographs	8:1491
Photocopying	8:1499
Radio	8:1626
Satellite	9:1707
Telegraph	10:1863
Telephone	10:1866
Television	10:1875
Video recording	10:1968

Computer science

Artificial intelligence	1:188
Automation	2:242
CAD/CAM	2:369
Calculator	2:370
Cellular/digital technology	3:439
Compact disc	3:531
Computer, analog	3:546
Computer, digital	3:547
Computer software	3:549
Internet	6:1123
Mass production	7:1236
Robotics	9:1690
Virtual reality	10:1969

Cosmology

Astrophysics	1:207
Big Bang theory	2:273
Cosmology	3:574
Galaxy	5:941
Space	9:1776

Cryogenics

Cryobiology	3:593
Cryogenics	3:595

Dentistry

Dentistry	4:626
Fluoridation	5:889

Ecology/Environmental science

Acid rain	1:9
Alternative energy sources	1:111
Biodegradable	2:280
Biodiversity	2:281

> Entries by Scientific Field

Bioenergy	**2**:284
Biome	**2**:293
Biosphere	**2**:304
Carbon cycle	**2**:389
Composting	**3**:539
DDT (dichlorodiphenyl-trichloroethane)	**4**:619
Desert	**4**:634
Dioxin	**4**:667
Drift net	**4**:680
Drought	**4**:682
Ecology	**4**:725
Ecosystem	**4**:728
Endangered species	**5**:793
Environmental ethics	**5**:807
Erosion	**5**:820
Eutrophication	**5**:828
Food web and food chain	**5**:894
Forestry	**5**:901
Forests	**5**:907
Gaia hypothesis	**5**:935
Greenhouse effect	**5**:1016
Hydrologic cycle	**6**:1071
Indicator species	**6**:1090
Nitrogen cycle	**7**:1342
Oil spills	**7**:1422
Organic farming	**7**:1431
Paleoecology	**8**:1457
Pollution	**8**:1549
Pollution control	**8**:1558
Rain forest	**8**:1641
Recycling	**9**:1650
Succession	**10**:1837
Waste management	**10**:2003
Wetlands	**10**:2024

Electrical engineering

Antenna	**1**:153
Battery	**2**:268
Cathode	**3**:415
Cathode-ray tube	**3**:417
Cell, electrochemical	**3**:436
Compact disc	**3**:531
Diode	**4**:665
Electric arc	**4**:734
Electric current	**4**:737
Electricity	**4**:741
Electric motor	**4**:747
Electrocardiogram	**4**:751
Electromagnetic field	**4**:758
Electromagnetic induction	**4**:760
Electromagnetism	**4**:766
Electronics	**4**:773
Fluorescent light	**5**:886
Generator	**5**:962
Incandescent light	**6**:1087
Integrated circuit	**6**:1106
LED (light-emitting diode)	**6**:1176
Magnetic recording/audiocassette	**6**:1209
Radar	**8**:1613
Radio	**8**:1626
Superconductor	**10**:1849
Telegraph	**10**:1863
Telephone	**10**:1866
Television	**10**:1875
Transformer	**10**:1908
Transistor	**10**:1910
Ultrasonics	**10**:1941
Video recording	**10**:1968

Electronics

Antenna	**1**:153
Battery	**2**:268
Cathode	**3**:415
Cathode-ray tube	**3**:417
Cell, electrochemical	**3**:436
Compact disc	**3**:531
Diode	**4**:665
Electric arc	**4**:734
Electric current	**4**:737
Electricity	**4**:741
Electric motor	**4**:747

Entries by Scientific Field

Electromagnetic field	4:758
Electromagnetic induction	4:760
Electronics	4:773
Generator	5:962
Integrated circuit	6:1106
LED (light-emitting diode)	6:1176
Magnetic recording/ audiocassette	6:1209
Radar	8:1613
Radio	8:1626
Superconductor	10:1849
Telephone	10:1866
Television	10:1875
Transformer	10:1908
Transistor	10:1910
Ultrasonics	10:1941
Video recording	10:1968

Embryology

Embryo and embryonic development	4:785
Fertilization	5:867
Reproduction	9:1664
Reproductive system	9:1667

Engineering

Aerodynamics	1:39
Aircraft	1:74
Antenna	1:153
Automation	2:242
Automobile	2:245
Balloon	1:261
Battery	2:268
Bridges	2:354
Canal	2:376
Cathode	3:415
Cathode-ray tube	3:417
Cell, electrochemical	3:436
Compact disc	3:531
Dam	4:611
Diesel engine	4:646
Diode	4:665
Electric arc	4:734
Electric current	4:737
Electric motor	4:747
Electricity	4:741
Electrocardiogram	4:751
Electromagnetic field	4:758
Electromagnetic induction	4:760
Electromagnetism	4:766
Electronics	4:773
Engineering	5:805
Fluorescent light	5:886
Generator	5:962
Incandescent light	6:1087
Integrated circuit	6:1106
Internal-combustion engine	6:1117
Jet engine	6:1143
LED (light-emitting diode)	6:1176
Lock	6:1192
Machines, simple	6:1203
Magnetic recording/ audiocassette	6:1209
Mass production	7:1236
Radar	8:1613
Radio	8:1626
Steam engine	9:1817
Submarine	10:1834
Superconductor	10:1849
Telegraph	10:1863
Telephone	10:1866
Television	10:1875
Transformer	10:1908
Transistor	10:1910
Ultrasonics	10:1941
Video recording	10:1968

Entomology

Arachnids	1:168
Arthropods	1:183

Butterflies	2:364	**General science**		*Entries by Scientific Field*
Cockroaches	3:505	Alchemy	1:82	
Insects	6:1103	Chaos theory	3:451	
Invertebrates	6:1133	Metric system	7:1265	
Metamorphosis	7:1259	Scientific method	9:1722	
		Units and standards	10:1948	

Epidemiology

Biological warfare	2:287
Disease	4:669
Ebola virus	4:717
Plague	8:1518
Poliomyelitis	8:1546
Sexually transmitted diseases	9:1735
Vaccine	10:1957

Genetic engineering

Biological warfare	2:287
Biotechnology	2:309
Clone and cloning	3:484
Genetic engineering	5:973

Evolutionary biology

Adaptation	1:26
Evolution	5:832
Human evolution	6:1054
Mendelian laws of inheritance	7:1246

Genetics

Biotechnology	2:309
Birth defects	2:319
Cancer	2:379
Carcinogen	2:406
Chromosome	3:472
Clone and cloning	3:484
Genetic disorders	5:966
Genetic engineering	5:973
Genetics	5:980
Human Genome Project	6:1060
Mendelian laws of inheritance	7:1246
Mutation	7:1314
Nucleic acid	7:1387

Food science

Brewing	2:352
Cyclamate	3:608
Food preservation	5:890
Nutrition	7:1399

Forensic science

Forensic science	5:898

Geochemistry

Coal	3:492
Earth (planet)	4:698
Earth science	4:707
Earth's interior	4:708
Glacier	5:1000
Minerals	7:1273
Rocks	9:1701
Soil	9:1758

Forestry

Forestry	5:901
Forests	5:907
Rain forest	8:1641
Tree	10:1927

Entries by Scientific Field

Geography

Africa	**1**:49
Antarctica	**1**:147
Asia	**1**:194
Australia	**2**:238
Biome	**2**:293
Cartography	**2**:410
Coast and beach	**3**:498
Desert	**4**:634
Europe	**5**:823
Geologic map	**5**:986
Island	**6**:1137
Lake	**6**:1159
Mountain	**7**:1301
North America	**7**:1352
River	**9**:1685
South America	**9**:1772

Geology

Catastrophism	**3**:415
Cave	**3**:420
Coal	**3**:492
Coast and beach	**3**:498
Continental margin	**3**:560
Dating techniques	**4**:616
Desert	**4**:634
Earthquake	**4**:702
Earth science	**4**:707
Earth's interior	**4**:708
Erosion	**5**:820
Fault	**5**:855
Geologic map	**5**:986
Geologic time	**5**:990
Geology	**5**:993
Glacier	**5**:1000
Hydrologic cycle	**6**:1071
Ice ages	**6**:1075
Iceberg	**6**:1078
Industrial minerals	**6**:1092
Island	**6**:1137
Lake	**6**:1159
Minerals	**7**:1273
Mining	**7**:1278
Mountain	**7**:1301
Natural gas	**7**:1319
Oil drilling	**7**:1418
Oil spills	**7**:1422
Petroleum	**8**:1492
Plate tectonics	**8**:1534
River	**9**:1685
Rocks	**9**:1701
Soil	**9**:1758
Uniformitarianism	**10**:1946
Volcano	**10**:1992
Water	**10**:2010

Geophysics

Earth (planet)	**4**:698
Earth science	**4**:707
Fault	**5**:855
Plate tectonics	**8**:1534

Gerontology

Aging and death	**1**:59
Alzheimer's disease	**1**:126
Arthritis	**1**:181
Dementia	**4**:622
Gerontology	**5**:999

Gynecology

Contraception	**3**:562
Fertilization	**5**:867
Gynecology	**5**:1022
Puberty	**8**:1599
Reproduction	**9**:1664

Health/Medicine

Acetylsalicylic acid	**1**:6
Addiction	**1**:32
Attention-deficit hyperactivity disorder (ADHD)	**2**:237

Entries by Scientific Field

Depression	**4:**630	Hallucinogens	**6:**1027
AIDS (acquired immunod-eficiency syndrome)	**1:**70	Immune system	**6:**1082
		Legionnaire's disease	**6:**1179
Alcoholism	**1:**85	Lipids	**6:**1191
Allergy	**1:**106	Malnutrition	**6:**1216
Alternative medicine	**1:**118	Marijuana	**6:**1224
Alzheimer's disease	**1:**126	Multiple personality disorder	**7:**1305
Amino acid	**1:**130		
Anesthesia	**1:**142	Nuclear medicine	**7:**1372
Antibiotics	**1:**155	Nutrition	**7:**1399
Antiseptics	**1:**164	Obsession	**7:**1405
Arthritis	**1:**181	Orthopedics	**7:**1434
Asthma	**1:**204	Parasites	**8:**1467
Attention-deficit hyperactivity disorder (ADHD)	**2:**237	Phobia	**8:**1497
		Physical therapy	**8:**1511
Birth defects	**2:**319	Plague	**8:**1518
Blood supply	**2:**330	Plastic surgery	**8:**1527
Burn	**2:**361	Poliomyelitis	**8:**1546
Carcinogen	**2:**406	Prosthetics	**8:**1579
Carpal tunnel syndrome	**2:**408	Protease inhibitor	**8:**1583
Cholesterol	**3:**469	Psychiatry	**8:**1592
Cigarette smoke	**3:**476	Psychology	**8:**1594
Cocaine	**3:**501	Psychosis	**8:**1596
Contraception	**3:**562	Puberty	**8:**1599
Dementia	**4:**622	Radial keratotomy	**8:**1615
Dentistry	**4:**626	Radiology	**8:**1637
Depression	**4:**630	Rh factor	**9:**1683
Diabetes mellitus	**4:**638	Schizophrenia	**9:**1716
Diagnosis	**4:**640	Sexually transmitted diseases	**9:**1735
Dialysis	**4:**644		
Disease	**4:**669	Sleep and sleep disorders	**9:**1745
Dyslexia	**4:**690	Stress	**9:**1826
Eating disorders	**4:**711	Sudden infant death syndrome (SIDS)	**10:**1840
Ebola virus	**4:**717		
Electrocardiogram	**4:**751	Surgery	**10:**1855
Fluoridation	**5:**889	Tranquilizers	**10:**1905
Food preservation	**5:**890	Transplant, surgical	**10:**1923
Genetic disorders	**5:**966	Tumor	**10:**1934
Genetic engineering	**5:**973	Vaccine	**10:**1957
Genetics	**5:**980	Virus	**10:**1974
Gerontology	**5:**999	Vitamin	**10:**1981
Gynecology	**5:**1022	Vivisection	**10:**1989

Entries by Scientific Field

Horticulture

Horticulture	**6:**1053
Plant	**8:**1522
Seed	**9:**1729
Tree	**10:**1927

Immunology

Allergy	**1:**106
Antibiotics	**1:**155
Antibody and antigen	**1:**159
Immune system	**6:**1082
Vaccine	**10:**1957

Marine biology

Algae	**1:**91
Amphibians	**1:**134
Cetaceans	**3:**448
Coral	**3:**566
Crustaceans	**3:**590
Endangered species	**5:**793
Fish	**5:**875
Mammals	**6:**1222
Mollusks	**7:**1288
Ocean zones	**7:**1414
Plankton	**8:**1520
Sponges	**9:**1799
Vertebrates	**10:**1967

Materials science

Abrasives	**1:**2
Adhesives	**1:**37
Aerosols	**1:**43
Alcohols	**1:**88
Alkaline earth metals	**1:**102
Alloy	**1:**110
Aluminum family	**1:**122
Artificial fibers	**1:**186
Asbestos	**1:**191
Biodegradable	**2:**280
Carbon family	**2:**395
Ceramic	**3:**447
Composite materials	**3:**536
Dyes and pigments	**4:**686
Electrical conductivity	**4:**731
Electrolysis	**4:**755
Expansion, thermal	**5:**842
Fiber optics	**5:**870
Glass	**5:**1004
Halogens	**6:**1030
Hand tools	**6:**1036
Hydrogen	**6:**1068
Industrial minerals	**6:**1092
Minerals	**7:**1273
Nitrogen family	**7:**1344
Oxygen family	**7:**1442
Plastics	**8:**1532
Polymer	**8:**1563
Soaps and detergents	**9:**1756
Superconductor	**10:**1849
Transition elements	**10:**1913

Mathematics

Abacus	**1:**1
Algebra	**1:**97
Arithmetic	**1:**177
Boolean algebra	**2:**333
Calculus	**2:**371
Chaos theory	**3:**451
Circle	**3:**478
Complex numbers	**3:**534
Correlation	**3:**569
Fractal	**5:**921
Fraction, common	**5:**923
Function	**5:**927
Game theory	**5:**945
Geometry	**5:**995
Graphs and graphing	**5:**1009
Imaginary number	**6:**1081
Logarithm	**6:**1195
Mathematics	**7:**1241

Multiplication	7:1307	El Niño	4:782	
Natural numbers	7:1321	Global climate	5:1006	
Number theory	7:1393	Monsoon	7:1291	
Numeration systems	7:1395	Ozone	7:1450	
Polygon	8:1562	Storm surge	9:1823	
Probability theory	8:1575	Thunderstorm	10:1887	
Proof (mathematics)	8:1578	Tornado	10:1900	
Pythagorean theorem	8:1601	Weather	10:2017	
Set theory	9:1733	Weather forecasting	10:2020	
Statistics	9:1810	Wind	10:2028	
Symbolic logic	10:1859			
Topology	10:1897			
Trigonometry	10:1931			
Zero	10:2047			

Entries by Scientific Field

Metallurgy

Alkali metals	1:99
Alkaline earth metals	1:102
Alloy	1:110
Aluminum family	1:122
Carbon family	2:395
Composite materials	3:536
Industrial minerals	6:1092
Minerals	7:1273
Mining	7:1278
Precious metals	8:1566
Transition elements	10:1913

Meteorology

Air masses and fronts	1:80
Atmosphere, composition and structure	2:211
Atmosphere observation	2:215
Atmospheric circulation	2:218
Atmospheric optical effects	2:221
Atmospheric pressure	2:225
Barometer	2:265
Clouds	3:490
Cyclone and anticyclone	3:608
Drought	4:682

Microbiology

Algae	1:91
Amoeba	1:131
Antiseptics	1:164
Bacteria	2:253
Biodegradable	2:280
Biological warfare	2:287
Composting	3:539
Parasites	8:1467
Plankton	8:1520
Protozoa	8:1590
Yeast	10:2043

Mineralogy

Abrasives	1:2
Ceramic	3:447
Industrial minerals	6:1092
Minerals	7:1273
Mining	7:1278

Molecular biology

Amino acid	1:130
Antibody and antigen	1:159
Biochemistry	2:279
Birth defects	2:319
Chromosome	3:472
Clone and cloning	3:484
Enzyme	5:812
Genetic disorders	5:966

Entries by Scientific Field

Genetic engineering 5:973
Genetics 5:980
Hormones 6:1050
Human Genome Project 6:1060
Lipids 6:1191
Molecular biology 7:1283
Mutation 7:1314
Nucleic acid 7:1387
Proteins 8:1586

Mycology

Brewing 2:352
Fermentation 5:864
Fungi 5:930
Yeast 10:2043

Nutrition

Diabetes mellitus 4:638
Eating disorders 4:711
Food web and food chain 5:894
Malnutrition 6:1216
Nutrition 7:1399
Vitamin 10:1981

Obstetrics

Birth 2:315
Birth defects 2:319
Embryo and embryonic development 4:785

Oceanography

Continental margin 3:560
Currents, ocean 3:604
Ocean 7:1407
Oceanography 7:1411
Ocean zones 7:1414
Tides 10:1890

Oncology

Cancer 2:379
Disease 4:669
Tumor 10:1934

Ophthalmology

Eye 5:848
Lens 6:1184
Radial keratotomy 8:1615

Optics

Atmospheric optical effects 2:221
Compact disc 3:531
Diffraction 4:648
Eye 5:848
Fiber optics 5:870
Hologram and holography 6:1048
Laser 6:1166
LED (light-emitting diode) 6:1176
Lens 6:1184
Light 6:1185
Luminescence 6:1196
Photochemistry 8:1498
Photocopying 8:1499
Telescope 10:1869
Television 10:1875
Video recording 10:1968

Organic chemistry

Carbon family 2:395
Coal 3:492
Cyclamate 3:608
Dioxin 4:667
Fermentation 5:864
Hydrogen 6:1068
Hydrologic cycle 6:1071
Lipids 6:1191

Natural gas	**7:**1319	Bacteria	**2:**253	
Nitrogen cycle	**7:**1342	Biological warfare	**2:**287	
Nitrogen family	**7:**1344	Cancer	**2:**379	
Oil spills	**7:**1422	Dementia	**4:**622	
Organic chemistry	**7:**1428	Diabetes mellitus	**4:**638	
Oxygen family	**7:**1442	Diagnosis	**4:**640	
Ozone	**7:**1450	Dioxin	**4:**667	
Petroleum	**8:**1492	Disease	**4:**669	
Vitamin	**10:**1981	Ebola virus	**4:**717	
		Genetic disorders	**5:**966	

Orthopedics

Arthritis	**1:**181
Orthopedics	**7:**1434
Prosthetics	**8:**1579
Skeletal system	**9:**1739

Malnutrition	**6:**1216
Orthopedics	**7:**1434
Parasites	**8:**1467
Plague	**8:**1518
Poliomyelitis	**8:**1546
Sexually transmitted diseases	**9:**1735
Tumor	**10:**1934
Vaccine	**10:**1957
Virus	**10:**1974

Paleontology

Dating techniques	**4:**616
Dinosaur	**4:**658
Evolution	**5:**832
Fossil and fossilization	**5:**917
Human evolution	**6:**1054
Paleoecology	**8:**1457
Paleontology	**8:**1459

Pharmacology

Acetylsalicylic acid	**1:**6
Antibiotics	**1:**155
Antiseptics	**1:**164
Cocaine	**3:**501
Hallucinogens	**6:**1027
Marijuana	**6:**1224
Poisons and toxins	**8:**1542
Tranquilizers	**10:**1905

Parasitology

Amoeba	**1:**131
Disease	**4:**669
Fungi	**5:**930
Parasites	**8:**1467

Physics

Acceleration	**1:**4
Acoustics	**1:**17
Aerodynamics	**1:**39
Antiparticle	**1:**163
Astrophysics	**1:**207
Atom	**2:**226
Atomic mass	**2:**229
Atomic theory	**2:**232
Ballistics	**2:**260

Pathology

AIDS (acquired immunodeficiency syndrome)	**1:**70
Alzheimer's disease	**1:**126
Arthritis	**1:**181
Asthma	**1:**204
Attention-deficit hyperactivity disorder (ADHD)	**2:**237

Entries by Scientific Field

Entries by Scientific Field

Battery	2:268	Gases, properties of	5:959
Biophysics	2:302	Generator	5:962
Buoyancy	2:360	Gravity and gravitation	5:1012
Calorie	2:375	Gyroscope	5:1024
Cathode	3:415	Half-life	6:1027
Cathode-ray tube	3:417	Heat	6:1043
Celestial mechanics	3:423	Hologram and holography	6:1048
Cell, electrochemical	3:436	Incandescent light	6:1087
Chaos theory	3:451	Integrated circuit	6:1106
Color	3:518	Interference	6:1112
Combustion	3:522	Interferometry	6:1114
Conservation laws	3:554	Ionization	6:1135
Coulomb	3:579	Isotope	6:1141
Cryogenics	3:595	Laser	6:1166
Dating techniques	4:616	Laws of motion	6:1169
Density	4:624	LED (light-emitting diode)	6:1176
Diffraction	4:648	Lens	6:1184
Diode	4:665	Light	6:1185
Doppler effect	4:677	Luminescence	6:1196
Echolocation	4:720	Magnetic recording/ audiocassette	6:1209
Elasticity	4:730	Magnetism	6:1212
Electrical conductivity	4:731	Mass	7:1235
Electric arc	4:734	Mass spectrometry	7:1239
Electric current	4:737	Matter, states of	7:1243
Electricity	4:741	Microwave communication	7:1268
Electric motor	4:747	Molecule	7:1285
Electrolysis	4:755	Momentum	7:1290
Electromagnetic field	4:758	Nuclear fission	7:1361
Electromagnetic induction	4:760	Nuclear fusion	7:1366
Electromagnetic spectrum	4:763	Nuclear medicine	7:1372
Electromagnetism	4:766	Nuclear power	7:1374
Electron	4:768	Nuclear weapons	7:1381
Electronics	4:773	Particle accelerators	8:1475
Energy	5:801	Periodic function	8:1485
Evaporation	5:831	Photochemistry	8:1498
Expansion, thermal	5:842	Photoelectric effect	8:1502
Fiber optics	5:870	Physics	8:1513
Fluid dynamics	5:882	Pressure	8:1570
Fluorescent light	5:886	Quantum mechanics	8:1607
Frequency	5:925	Radar	8:1613
Friction	5:926	Radiation	8:1619
Gases, liquefaction of	5:955		

Entries by Scientific Field

Radiation exposure	8:1621
Radio	8:1626
Radioactive tracers	8:1629
Radioactivity	8:1630
Radiology	8:1637
Relativity, theory of	9:1659
Sonar	9:1770
Spectroscopy	9:1792
Spectrum	9:1794
Subatomic particles	10:1829
Superconductor	10:1849
Telegraph	10:1863
Telephone	10:1866
Television	10:1875
Temperature	10:1879
Thermal expansion	5:842
Thermodynamics	10:1885
Time	10:1894
Transformer	10:1908
Transistor	10:1910
Tunneling	10:1937
Ultrasonics	10:1941
Vacuum	10:1960
Vacuum tube	10:1961
Video recording	10:1968
Virtual reality	10:1969
Volume	10:1999
Wave motion	10:2014
X ray	10:2033

Primatology

Animal	1:145
Endangered species	5:793
Mammals	6:1222
Primates	8:1571
Vertebrates	10:1967

Psychiatry/Psychology

Addiction	1:32
Alcoholism	1:85
Attention-deficit hyperactivity disorder (ADHD)	2:237
Behavior	2:270
Cognition	3:511
Depression	4:630
Eating disorders	4:711
Multiple personality disorder	7:1305
Obsession	7:1405
Perception	8:1482
Phobia	8:1497
Psychiatry	8:1592
Psychology	8:1594
Psychosis	8:1596
Reinforcement, positive and negative	9:1657
Savant	9:1712
Schizophrenia	9:1716
Sleep and sleep disorders	9:1745
Stress	9:1826

Radiology

Nuclear medicine	7:1372
Radioactive tracers	8:1629
Radiology	8:1637
Ultrasonics	10:1941
X ray	10:2033

Robotics

Automation	2:242
Mass production	7:1236
Robotics	9:1690

Seismology

Earthquake	4:702
Volcano	10:1992

Sociology

Adaptation	1:26
Aging and death	1:59

Entries by Scientific Field

Alcoholism	1:85
Behavior	2:270
Gerontology	5:999
Migration (animals)	7:1271

Technology

Abrasives	1:2
Adhesives	1:37
Aerosols	1:43
Aircraft	1:74
Alloy	1:110
Alternative energy sources	1:111
Antenna	1:153
Artificial fibers	1:186
Artificial intelligence	1:188
Asbestos	1:191
Automation	2:242
Automobile	2:245
Balloon	1:261
Battery	2:268
Biotechnology	2:309
Brewing	2:352
Bridges	2:354
CAD/CAM	2:369
Calculator	2:370
Canal	2:376
Cathode	3:415
Cathode-ray tube	3:417
Cell, electrochemical	3:436
Cellular/digital technology	3:439
Centrifuge	3:445
Ceramic	3:447
Compact disc	3:531
Computer, analog	3:546
Computer, digital	3:547
Computer software	3:549
Cybernetics	3:605
Dam	4:611
Diesel engine	4:646
Diode	4:665
DVD technology	4:684
Dyes and pigments	4:686
Fiber optics	5:870
Fluorescent light	5:886
Food preservation	5:890
Forensic science	5:898
Generator	5:962
Glass	5:1004
Hand tools	6:1036
Hologram and holography	6:1048
Incandescent light	6:1087
Industrial Revolution	6:1097
Integrated circuit	6:1106
Internal-combustion engine	6:1117
Internet	6:1123
Jet engine	6:1143
Laser	6:1166
LED (light-emitting diode)	6:1176
Lens	6:1184
Lock	6:1192
Machines, simple	6:1203
Magnetic recording/audiocassette	6:1209
Mass production	7:1236
Mass spectrometry	7:1239
Microwave communication	7:1268
Paper	8:1462
Photocopying	8:1499
Plastics	8:1532
Polymer	8:1563
Prosthetics	8:1579
Radar	8:1613
Radio	8:1626
Robotics	9:1690
Rockets and missiles	9:1693
Soaps and detergents	9:1756
Sonar	9:1770
Space station, international	9:1788
Steam engine	9:1817
Submarine	10:1834
Superconductor	10:1849
Telegraph	10:1863
Telephone	10:1866

Television	**10:**1875
Transformer	**10:**1908
Transistor	**10:**1910
Vacuum tube	**10:**1961
Video recording	**10:**1968
Virtual reality	**10:**1969

Virology

AIDS (acquired immuno-deficiency syndrome)	**1:**70
Disease	**4:**669
Ebola virus	**4:**717
Plague	**8:**1518
Poliomyelitis	**8:**1546
Sexually transmitted diseases	**9:**1735
Vaccine	**10:**1957
Virus	**10:**1974

Weaponry

Ballistics	**2:**260
Biological warfare	**2:**287
Chemical warfare	**3:**457
Forensic science	**5:**898
Nuclear weapons	**7:**1381
Radar	**8:**1613
Rockets and missiles	**9:**1693

Wildlife conservation

Biodiversity	**2:**281
Biome	**2:**293
Biosphere	**2:**304
Drift net	**4:**680
Ecology	**4:**725
Ecosystem	**4:**728
Endangered species	**5:**793
Forestry	**5:**901
Gaia hypothesis	**5:**935
Wetlands	**10:**2024

Zoology

Amphibians	**1:**134
Animal	**1:**145
Arachnids	**1:**168
Arthropods	**1:**183
Behavior	**2:**270
Birds	**2:**312
Butterflies	**2:**364
Canines	**2:**382
Cetaceans	**3:**448
Cockroaches	**3:**505
Coelacanth	**3:**508
Coral	**3:**566
Crustaceans	**3:**590
Dinosaur	**4:**658
Echolocation	**4:**720
Endangered species	**5:**793
Felines	**5:**855
Fish	**5:**875
Hibernation	**6:**1046
Indicator species	**6:**1090
Insects	**6:**1103
Invertebrates	**6:**1133
Kangaroos and wallabies	**6:**1153
Mammals	**6:**1222
Metamorphosis	**7:**1259
Migration (animals)	**7:**1271
Mollusks	**7:**1288
Plankton	**8:**1520
Primates	**8:**1571
Reptiles	**9:**1670
Snakes	**9:**1752
Sponges	**9:**1799
Vertebrates	**10:**1967

Entries by Scientific Field

Catalyst and catalysis

Catalysis (pronounced cat-AL-uh-sis) is the process by which some substance is added to a reaction in order to make the reaction occur more quickly. The substance that is added to produce this result is the catalyst (pronounced CAT-uh-list).

You are probably familiar with the catalytic convertor, a device used in car exhaust systems to remove gases that cause air pollution. The catalytic convertor gets its name from the fact that certain metals (the catalysts) inside the device cause exhaust gases to break down. For example, when potentially dangerous nitrogen(II) oxide passes through a catalytic convertor, platinum and rhodium catalysts cause the oxide to break down into harmless nitrogen and oxygen. Nitrogen(II) oxide will break down into nitrogen and oxygen even without the presence of platinum and rhodium. However, that process takes place over hours, days, or weeks under natural circumstances. By that time, the dangerous gas is already in the atmosphere. In the catalytic convertor, the breakdown of nitrogen(II) oxide takes place within a matter of seconds.

History

Humans have known about catalysis for many centuries, even though they knew nothing about the chemical process that was involved. The making of soap, the fermentation of wine to vinegar, and the leavening of bread are all processes involving catalysis. One of the first formal experiments on catalysis occurred in 1812. Russian chemist Gottlieb Sigismund Constantin Kirchhof (1764–1833) studied the behavior of starch in boiling water. Under most circumstances, Kirchhof found, no change occurred

Catalyst and catalysis

when starch was simply boiled in water. But adding just a few drops of concentrated sulfuric acid to the boiling water had a profound effect on the starch. In very little time, the starch broke down to form the simple sugar known as glucose. When Kirchhof found that the sulfuric acid remained unchanged at the completion of the experiment, he concluded that it had simply played a helping role in the conversion of starch to sugar.

The name catalysis was actually proposed in 1835 by Swedish chemist Jöns Jakob Berzelius (1779–1848). The word comes from two Greek terms, *kata* (for "down") and *lyein* (for "loosen"). Berzelius used the term to emphasize that the process loosens the bonds by which chemical compounds are held together.

Types of catalysis

Catalysis reactions are usually categorized as either homogeneous or heterogeneous reactions. A homogeneous catalysis reaction is one in which both the catalyst and the substances on which it works are all in the same phase (solid, liquid, or gas). The reaction studied by Kirchhof is an example of a homogeneous catalysis. Both the sulfuric acid and the starch were in the same phase—a water solution—during the reaction.

A heterogeneous catalysis reaction is one in which the catalyst is in a different phase from the substances on which it acts. In a catalytic convertor, for example, the catalyst is a solid, usually a precious metal such as platinum or rhodium. The substances on which the catalyst acts, however, are gases, such as nitrogen(II) oxide and other gaseous products of combustion.

Some of the most interesting and important catalysts are those that occur in living systems: the enzymes. All of the reactions that take place within living bodies occur naturally, whether or not a catalyst is present. But they take place so slowly on their own that they are of no value to the survival of an organism. For example, if you place a sugar cube in a glass of water, it eventually breaks down into simpler molecules with the release of energy. But that process might take years. A person who ate a sugar cube and had to wait that long for the energy to be released in the body would die.

Fortunately, our bodies contain catalysts (enzymes) that speed up such reactions. They make it possible for the energy stored in sugar molecules to be released in a matter of minutes.

Industrial applications

Today catalysts are used in untold numbers of industrial processes. For example, the commercially important gas ammonia is produced by combining nitrogen gas and hydrogen gas at a high temperature and pressure in the presence of a catalyst such as powdered iron. In the absence of the catalyst, the reaction between nitrogen and hydrogen would, for all practical purposes, not occur. In its presence, the reaction occurs quickly enough to produce ammonia gas in large quantities.

[*See also* **Enzyme; Reaction, chemical**]

Catastrophism

In geology, catastrophism is the belief that Earth's features—including mountains, valleys, and lakes—were created suddenly as a result of great catastrophes, such as floods or earthquakes. This is the opposite of uniformitarianism, the view held by many present-day scientists that Earth's features developed gradually over long periods of time.

Catastrophism developed in the seventeenth and eighteenth centuries when tradition and even the law forced scientists to use the Bible as a scientific document. Theologians (religious scholars) of the time believed Earth was only about 6,000 years old (current scientific research estimates Earth to be 4.5 billion years old). Based on this thinking and the supernatural events described in the book of Genesis in the Bible, geologists concluded that fossils of ocean-dwelling organisms were found on mountain tops because of Noah's flood. The receding flood waters also carved valleys, pooled in lakes, and deposited huge boulders far from their sources.

Over the next 200 years, as geologists developed more scientific explanations for natural history, catastrophism was abandoned. Since the late 1970s, however, another form of catastrophism has arisen with the idea that large objects from space periodically collide with Earth, destroying life. Scientists speculate that when these objects strike, they clog the atmosphere with sunlight-blocking dust and gases. One theory holds that the most famous of these collisions killed off the dinosaurs roughly 65 million years ago.

[*See also* **Uniformitarianism**]

Cathode

A cathode is one of the two electrodes used either in a vacuum tube or in an electrochemical cell. An electrode is the part (pole) of a vacuum tube or cell through which electricity moves into or out of the system. The other pole of the system is referred to as the anode.

Cathode

Vacuum tubes

A vacuum tube is a hollow glass cylinder from which as much air as possible has been removed. In a vacuum tube, the cathode is the negative electrode. It has more electrons on its surface than does the other electrode, the anode. Electrons can accumulate on the surface of a cathode for various reasons. For example, in some vacuum tubes, the cathode is heated to a high temperature to remove electrons from atoms that make up the cathode. The free electrons are then able to travel from the cathode to the anode. These streams of electrons are known as cathode rays, and the tubes in which they are produced are called cathode-ray tubes (CRTs). CRTs are widely used as oscilloscopes (which measure changes in electrical voltage over time), television tubes, and computer monitors.

Electrochemical cells

Electrochemical cells are devices for turning chemical energy into electrical energy or, alternatively, changing electrical energy into chemical energy. Electrochemical cells are of two types: voltaic cells (also called galvanic cells) and electrolytic cells. In a voltaic or galvanic cell, electrical energy is produced as the result of a chemical reaction between two different metals immersed in a (usually) water solution. The differing tendency of the two metals to gain and lose electrons causes an electric current to flow through an external wire connecting the two metals. The cathode is defined in a system of this type as the metal at which electrons are being taken up from the external wire. In contrast, the anode is the point at which electrons are being given up to the external wire.

Practical applications

Cathodes are used in many practical applications. For example, electroplating is a process by which a layer of pure metal can be deposited on a base used as the cathode. Suppose that a spoon composed of iron is made the cathode in an electrochemical cell that also contains an anode made of silver metal and a solution of silver nitrate. In this cell, silver atoms lose electrons that travel through an external circuit to the cathode. At the cathode, the electrons combine with silver ions in the solution to form silver atoms. These silver atoms plate out on the surface of the iron spoon, giving it a coating of silver metal. The plain iron spoon soon develops a shiny silver surface. It looks more attractive and is less likely to rust than the original iron spoon.

[See also **Cathode-ray tube; Cell, electrochemical; Vacuum tube**]

Cathode-ray tube

A cathode-ray tube is a device that uses a beam of electrons in order to produce an image on a screen. Cathode-ray tubes, also known commonly as CRTs, are widely used in a number of electrical devices such as computer screens, television sets, radar screens, and oscilloscopes used for scientific and medical purposes.

Any cathode-ray tube consists of five major parts: an envelope or container, an electron gun, a focusing system, a deflection system, and a display screen.

Envelope or container

Most people have seen a cathode-ray tube or pictures of one. The picture tube in a television set is perhaps the most familiar form of a cathode-ray tube. The outer shell that gives a picture tube its characteristic shape is called the envelope of a cathode-ray tube. The envelope is most commonly made of glass, although tubes of metal and ceramic can also be used for special purposes. The glass cathode-ray tube consists of a cylindrical portion that holds the electron gun and the focusing and deflection systems. At the end of the cylindrical portion farthest from the electron gun, the tube widens out to form a conical shape. At the flat wide end of the cone is the display screen.

Air is pumped out of the cathode-ray tube to produce a vacuum with a pressure in the range of 10^{-2} to 10^{-6} pascal (units of pressure), the exact value depending on the use to which the tube will be put. A vacuum is necessary to prevent electrons produced in the CRT from colliding with atoms and molecules within the tube.

Electron gun

An electron gun consists of three major parts. The first is the cathode—a piece of metal which, when heated, gives off electrons. One of the most common cathodes in use is made of cesium metal, a member of the alkali family that loses electrons very easily. When a cesium cathode is heated to a temperature of about 1750°F (approximately 825°C), it begins to release a stream of electrons. These electrons are then accelerated by an anode (a positively charged electrode) placed a short distance away from the cathode. As the electrons are accelerated, they pass through a small hole in the anode into the center of the cathode-ray tube.

The intensity of the electron beam entering the anode is controlled by a grid. The grid may consist of a cylindrical piece of metal to which

Cathode-ray tube

a variable electrical charge can be applied. The amount of charge placed on the control grid determines the intensity of the electron beam that passes through it.

Focusing and deflection systems

Under normal circumstances, an electron beam produced by an electron gun tends to spread out, forming a cone-shaped beam. However, the beam that strikes the display screen must be pencil-thin and clearly defined. In order to form the electron beam into the correct shape, an electrical or magnetic lens must be added to the CRT. The lens is similar to an optical lens, like the lens in a pair of glasses. The electrical or mag-

A 1962 photo of a wire-caged cathode-ray tube. The glass vacuum chamber encloses an electron gun. (Reproduced courtesy of the Library of Congress.)

Oscilloscope

An especially useful application of the cathode-ray tube is an oscilloscope. An oscilloscope measures changes in electrical voltage over time. The plates that deflect the electron beam in a vertical direction are attached to some source of voltage. (For example, they can be connected directly to an electric circuit.) The plates deflecting the electron beam in a horizontal direction are attached to some sort of a clock mechanism.

Wired in this way, the oscilloscope shows the change in voltage in a circuit over time. This change shows up as a wavy line on a screen. As voltage increases, the line moves upward. As it decreases, the line moves downward.

One application of the oscilloscope is troubleshooting an electric circuit. An observer can notice immediately if a problem has developed within a circuit. For example, circuits can be damaged if unusually large voltages develop very quickly. If a circuit is being monitored on an oscilloscope, such voltage surges can be detected immediately. Oscilloscopes also have medical applications. They can be connected to electrodes attached to a person's skin. The electrodes measure very small voltage changes in the person's body. Such changes can be an indication of the general health of the person's nervous system.

---

netic lens shapes the flow of electrons that pass through it, just as a glass lens shapes the light rays passing through it.

The electron beam in a cathode-ray tube also has to be moved back and forth so that it can strike any part of the display screen. In general, two kinds of systems are available for controlling the path of the electron beam: one uses electrical charges and the other uses a magnetic field. In either case, two deflection systems are needed: one to move the electron beam in a horizontal direction and the other to move it in a vertical direction. In a standard television tube, the electron beam completely scans the display screen about 25 times every second.

Display screen

The actual conversion of electrical energy to light energy takes place on the display screen when electrons strike a material known as a phosphor. A phosphor is a chemical that glows when exposed to electri-

Cave

cal energy. A commonly used phosphor is the compound zinc sulfide. When pure zinc sulfide is struck by an electron beam, it gives off a greenish glow. The exact color given off by a phosphor also depends on the presence of small amounts of impurities. For example, zinc sulfide with silver metal as an impurity gives off a bluish glow, while zinc sulfide with copper metal as an impurity gives off a greenish glow.

The selection of phosphors to be used in a cathode-ray tube is very important. Many different phosphors are known, and each has special characteristics. For example, the phosphor known as yttrium oxide gives off a red glow when struck by electrons, and yttrium silicate gives off a purplish-blue glow.

The rate at which a phosphor responds to an electron beam is also of importance. In a color television set, for example, the glow produced by a phosphor has to last long enough, but not too long. Remember that the screen is being scanned 25 times every second. If the phosphor continues to glow too long, color will remain from the first scan when the second scan has begun, and the overall picture will become blurred. On the other hand, if the color from the first scan fades out before the second scan has begun, the screen will go blank briefly, making the picture flicker.

Cathode-ray tubes differ in their details of construction depending on the use to which they will be put. In an oscilloscope, for example, the electron beam has to be able to move about on the screen very quickly and with high precision, although it needs to display only one color. Factors such as size and durability are also more important in an oscilloscope than they might be in a home television set. In a commercial television set, on the other hand, color is obviously an important factor. In such a set, a combination of three electron guns is needed—one for each of the primary colors used in making the color picture.

Cave

A cave is a naturally occurring hollow area inside Earth. All caves are formed by some type of erosion process. The study of caves is called speleology (pronounced spee-lee-OL-o-gee). While some caves may be small hillside openings, others may consist of large chambers and interconnecting tunnels and mazes. Openings to the surface may be large gaping holes or small crevices.

Caves have served as shelter for people throughout history. Many religious traditions have regarded caves as sacred and have used them in rituals and ceremonies. Human remains, artifacts, sculptures, and draw-

> ### Words to Know
>
> **Speleology:** Scientific study of caves and their plant and animal life.
>
> **Stalactite:** Cylindrical or icicle-shaped mineral deposit projecting downward from the roof of a cave.
>
> **Stalagmite:** Cylindrical or upside down icicle-shaped mineral deposit projecting upward from the floor of a cave.

ings found in caves have aided archaeologists in learning about early humans. A cave discovered in southeastern France in 1994 contains wall paintings estimated to be more than 30,000 years old.

Cave formation

The most common, largest, and spectacular caves are solution caves. These caves are formed through the chemical interaction of air, soil, water, and rock. As water flows over and drains into Earth's surface, it mixes with carbon dioxide from the air and soil to form a mild solution of carbonic acid. Seeping through naturally occurring cracks and fissures in massive beds of limestone in bedrock (the solid rock that lies beneath the soil), the acidic water eats away at the rock, dissolving its minerals and carrying them off in a solution.

With continual water drainage, the fissures become established passageways. The passageways eventually enlarge and often connect, creating an underground drainage system. Sometimes ceilings fall and passageways collapse, creating new spaces and drainage routes. Over thousands, perhaps millions of years, these passages evolve into the caves we see today.

Several distinctive features in the landscape make cave terrain easy to identify. The most common is a rugged land surface, marred by sinkholes, circular depressions where the underlying rock has been dissolved away. Disappearing streams and natural bridges are also common clues. But entrances to solution caves are not always obvious, and their discovery is sometimes quite by accident.

Cave environment

A deep cave is completely dark, has a stable atmosphere, and has an almost constant temperature. The humidity in limestone caves is usu-

Cave

ally near 100 percent. Many caves contain unique life-forms, underground streams and lakes, and unusual mineral formations.

Water that makes its way to a cave ceiling hangs as a drop. The damp atmosphere in a cave reacts with that water, forcing the dissolved mineral out of the water solution. The crystalline material that most often remains is called calcite. Calcite deposited on the ceiling creates a hanging icicle-shaped formation called a stalactite (pronounced sta-LACK-tite). Calcite deposited on the floor of a cave builds up to create an upside down icicle-shaped formation called a stalagmite (pronounced

Stalactites and stalagmites in Harrison's Cave, Barbados. *(Reproduced courtesy of Carol DeKane Nagel.)*

sta-LAG-mite). Stalactites and stalagmites grow by only a fraction of an inch or centimeter a year. In time, two such formations often merge to form a stout floor-to-ceiling column.

Sometimes the water runs down the slope of the wall, and as the calcite is deposited, a low ridge forms. Subsequent drops of water follow the ridge, adding more calcite. Constant buildup of calcite in this fashion results in the formation of a wavy, folded sheet hanging from the ceiling called a curtain. Curtain formations often have streaks of various shades of off-white and brown.

Cave life

Three different groups of animals use or inhabit caves. Animals in the first group commonly use caves but depend on the outside world for survival. These include bats, birds, bears, and crickets. Those in the second group live their entire life cycle within a cave, generally near the entrance, but are also found living outside caves. Cockroaches, beetles, and millipedes are some examples of this second group. The last group comprises animals that are permanent deep cave dwellers. Because they often live in total darkness, these animals lack skin color and eyes. They rely on their sense of touch to get around. Examples of this group include fish, shrimp, crayfish, salamanders, worms, snails, insects, bacteria, fungi, and algae.

Celestial mechanics

Celestial mechanics is a branch of astronomy that studies the movement of bodies in outer space. Using a mathematical theory, it explains the observed motion of the planets and allows us to predict their future movements. It also comes into play when we launch a satellite into space and expect to direct its flight.

Early Greeks

Until English scientist and mathematician Isaac Newton (1642–1727) founded the science of celestial mechanics over 300 years ago, the movement of the planets regularly baffled astronomers or anyone who studied the heavens. This is because those bodies called planets, a word which comes from the Greek word for "wanderer," literally "wandered" about the sky in a seemingly unpredictable manner. To the early astronomers, the stars were fixed in the heavens and the Sun seemed to make the same

Celestial mechanics

> ### Words to Know
>
> **Copernican system:** Theory proposing that the Sun is at the center of the solar system and all planets, including Earth, revolve around it.
>
> **Epicycle:** A circle on which a planet moves and which has a center that is itself carried around at the same time on the rim of a larger circle.
>
> **Gravity:** Force of attraction between objects, the strength of which depends on the mass of each object and the distance between them. Also, the special acceleration of 9.81 meters per second per second exerted by the attraction of the mass of Earth on nearby objects.
>
> **Orbit:** The path followed by a body (such as a planet) in its travel around another body (such as the Sun).
>
> **Ptolemaic system:** Theory proposing that Earth is at the center of the solar system and the Sun, the Moon, and all the planets revolve around it.

regular journey every year, but the planets followed no such pattern. Their unpredictable behavior was especially frustrating to the ancients, and it was not until around A.D. 140 that Greek astronomer Claudius Ptolemaeus or Ptolemy provided some kind of order to this chaotic situation.

In what came to be known as the Ptolemaic (pronounced tahl-uh-MAY-ik) system, he placed Earth at the center of the universe and had the Sun revolving around it, along with all the other known planets. Ptolemy's system of predicting which planet would be where and when was described as the epicyclic theory because it was based on the notion of epicycles (an epicycle is a small circle whose center is on the rim of a larger circle). Since Ptolemy and most Greeks of his time believed that all planetary motion must be circular (which it is not), they had to keep adding more and more epicycles to their calculations to make their system work. Despite the fact that it was very complicated and difficult to use, his system was able generally to tell where the planets would be with some degree of accuracy. This is even more amazing when we realize that it was based on an entirely incorrect notion of the solar system (since it had the Earth and not the Sun at the center). Yet its ingenious use of the off-center epicycles, which were regularly adjusted, permitted the Ptolemaic system to approximate the irregular movements of the planets.

This, in turn, mostly accounts for the fact that his incorrect system survived and was used by every astronomer and astrologer for 1,400 years.

Copernican revolution

During the Middle Ages (period in European history usually dated from about 500 to 1450), the Ptolemaic system was dominant, and all educated people whether in Europe or in the Arabian world used it to explain the movement of the planets. By then, the system had the seven known planets riding more than 240 different epicycles. It was this level of complexity that led Spanish King Alfonso X (1221–1284), who was called "the Wise," to state that had he been present at the creation of the world, he would have suggested that God make a simpler planetary system. The clutter of all of this complexity was eventually done away with (although by no means immediately) when Polish astronomer Nicolaus Copernicus (1473–1543) offered what is called his heliocentric (pronounced hee-lee-oh-SEN-trik) theory in 1543.

Copernicus placed the Sun at the center of the solar system and made the planets (including Earth) orbit the Sun on eccentric circles (which are more egg-shaped than perfectly circular or round ones). The Copernican system took a long time to be adopted, mainly because it was actively condemned for over a century by the Catholic Church. The Church objected to the fact that his system took Earth out of its stationary center position and made it revolve instead around the Sun with all the other planets. Although Copernicus could explain certain phenomena—for example he correctly stated that the farther a planet lies from the Sun the slower it moves—his system still did not have a mathematical formula that could be used to explain and predict planetary movement.

Kepler's laws

By the time the Church was condemning the work of Italian astronomer and physicist Galileo Galilei (1564–1642), who defended the Copernican model of the solar system, German astronomer Johannes Kepler (1571–1630) had already published his three laws of planetary motion, which would lay the groundwork for all of modern astronomy. His first two laws were contained in his *Astronomia nova* (*The New Astronomy*), published in 1609, and his third was stated in his book *Harmonices mundi* (*Harmony of the World*), published in 1619. Basically, the laws state that the orbits of planets can be drawn as ellipses (elongated egg shapes) with the Sun always at one of their central points; that a planet moves faster the closer it is to the Sun (and slower the farther away it is); and, lastly,

Celestial mechanics

that it is possible to calculate a planet's relative distance from the Sun knowing its period of revolution. Kepler's laws about the planets and the Sun laid the groundwork for English physicist Isaac Newton to be able to go further and generalize about what might be called the physics of the universe—in other words, the mechanics of the heavens or celestial mechanics (celestial being another word for "the heavens").

Newtonian mechanics

Celestial mechanics is, therefore, Newtonian mechanics. Newton's greatness was in his ability to seek out and find a generalization or a single big idea that would explain the behavior of bodies in motion. Newton was able to do this with what is called his law of universal gravitation and his three laws of motion. The amazing thing about his achievement is that he discovered certain general principles that unified the heavens and Earth. He showed that all aspects of the natural world, near and far, were subject to the same laws of motion and gravitation, and that they could be demonstrated in mathematical terms within a single theory.

In 1687, Newton published his epic work, *Philosophiae naturalis principia mathematica* (*Mathematical Principles of Natural Philosophy*). In the first part of the book, Newton offers his three laws of motion. The first law is the principle of inertia, which says that a body stays at rest (or in motion) until an outside force acts upon it. His second law defines force as the product of how fast something is moving and how much matter (called mass) is in it. His third law says that for every action there is an equal and opposite reaction. It was from these laws that Newton arrived at his law of universal gravitation, which can be said to have founded the science of celestial mechanics.

Gravity as a universal force

The law of universal gravitation states that every particle of matter attracts every other particle with a force that is directly proportional (an equal ratio such as 1:1) to the product of the masses of the particles, and is inversely proportional (the opposite ratio) to the square of the distance between them. Although this may sound complicated, it actually simplified things because celestial mechanics now had an actual set of equations that could be used with the laws of motion to figure out how two bodies in space influenced and affected each other.

It was Newton's great achievement that he discovered gravity to be the force that holds the universe together. Gravity is a mutual attraction or a two-way street between bodies. That is, a stone falls to the ground mainly because Earth's gravity pulls it downward (since Earth's mass is much greater than that of the stone). But the stone also exerts its influence on Earth, although it is so tiny it has no effect. However, if the two bodies were closer in size, this two-way attraction would be more noticeable. We see this with Earth and the Moon. Earth's gravity holds the Moon in orbit around it, but just as Earth exerts a force on the Moon, so the Moon pulls upon Earth. We can demonstrate this by seeing how the free-flowing water of the oceans gets pulled toward the side of Earth that is facing the Moon (what we call high tide). The opposite side of Earth also experiences this same thing at the same time, as the ocean on that side also bulges away from Earth since the Moon's gravity pulls the solid body of Earth away from the water on Earth's distant side.

When the principles discovered by Newton are applied only to the movements of bodies in outer space, it is called celestial mechanics instead of just mechanics. Therefore, using Newton's laws, we can analyze the orbital movements of planets, comets, asteroids, and human-made

The unmanned Pioneer 10 spacecraft leaving our solar system. The laws of celestial mechanics allow scientists to determine the orbits of artificial satellites. *(Reproduced by permission of The Stock Market.)*

satellites and spacecraft, as well as the motions of stars and even galaxies. However, Newton's solution works best (and easiest) when there are only two bodies (like Earth and the Moon) involved. The situation becomes incredibly complicated when there are three or more separate forces acting on each other at once, and all these bodies are also moving at the same time. This means that each body is subject to small changes that are known as perturbations (pronounced pur-tur-BAY-shunz). These perturbations or small deviations do not change things very much in a short period of time, but over a very long period they may add up and make a considerable difference.

That is why today's celestial mechanics of complicated systems are really only very good approximations. However, computer advances have made quite a difference in the degree of accuracy achieved. Finally, with the beginning of the space age in 1957 when the first artificial satellite was launched, a new branch of celestial mechanics called astrodynamics was founded that considers the effects of rocket propulsion in putting an object into the proper orbit or extended flight path. Although our space activity has presented us with new and complicated problems of predicting the motion of bodies in space, it is still all based on the celestial mechanics laid out by Isaac Newton over three centuries ago.

[*See also* **Gravity and gravitation; Laws of motion; Orbit; Tides**]

Cell

The cell is the basic unit of a living organism. In multicellular organisms (organisms with more than one cell), a collection of cells that work together to perform similar functions is called a tissue. In the next higher level of organization, various tissues that perform coordinated functions form organs. Finally, organs that work together to perform general processes form body systems.

Types of cells

Multicellular organisms contain a vast array of highly specialized cells. Plants contain root cells, leaf cells, and stem cells. Humans have skin cells, nerve cells, and sex cells. Each kind of cell is structured to perform a highly specialized function. Often, examining a cell's structure reveals much about its function in the organism. For instance, certain cells in the small intestine have developed microvilli (hairs) that promote the absorption of foods. Nerve cells, or neurons, are another kind of specialized cell whose form reflects function. Nerve cells consist of a cell body and long attachments, called axons, that conduct nerve impulses. Dendrites are shorter attachments that receive nerve impulses.

Sensory cells are cells that detect information from the outside environment and transmit that information to the brain. Sensory cells often have unusual shapes and structures that contribute to their function. The rod cells in the retina of the eye, for instance, look like no other cell in the human body. Shaped like a rod, these cells have a light-sensitive region that contains numerous disks. Within each disk is embedded a special light-sensitive pigment that captures light. When the pigment receives light from the outside environment, nerve cells in the eye are triggered to send a nerve impulse in the brain. In this way, humans are able to detect light.

Cells, however, can also exist as single-celled organisms. The organisms called protists, for instance, are single-celled organisms. Examples of protists include the microscopic organism called *Paramecium* and the single-celled alga called *Chlamydomonas*.

Prokaryotes and eukaryotes. Two types of cells are recognized in living things: prokaryotes and eukaryotes. The word prokaryote literally means "before the nucleus." As the name suggests, prokaryotes are cells that have no distinct nucleus. Most prokaryotic organisms are single-celled, such as bacteria and algae.

The term eukaryote means "true nucleus." Eukaryotes have a distinct nucleus and distinct organelles. An organelle is a small structure that performs a specific set of functions within the eukaryotic cell. These organelles are held together by membranes. In addition to their lack of a nucleus, prokaryotes also lack these distinct organelles.

The structure and function of cells

The basic structure of all cells, whether prokaryote and eukaryote, is the same. All cells have an outer covering called a plasma membrane. The plasma membrane holds the cell together and permits the passage of substances into and out of the cell. With a few minor exceptions, plasma membranes are the same in prokaryotes and eukaryotes.

The interior of both kinds of cells is called the cytoplasm. Within the cytoplasm of eukaryotes are embedded the cellular organelles. As noted above, the cytoplasm of prokaryotes contains no organelles. Finally, both types of cells contain small structures called ribosomes. Ribosomes are the sites within cells where proteins are produced. (Proteins are large molecules that are essential to the structure and functioning of all living

Cell

> ### Words to Know
>
> **Cell wall:** A tough outer covering that overlies the plasma membrane of bacteria and plant cells.
>
> **Cilia:** Short projections that cover the surface of some cells and provide for movement.
>
> **Cytoplasm:** The semifluid substance of a cell containing organelles and enclosed by the cell membrane.
>
> **Cytoskeleton:** The network of filaments that provide structure and movement of a cell.
>
> **DNA (deoxyribonucleic acid):** The genetic material in the nucleus of cells that contains information for an organism's development.
>
> **Endoplasmic reticulum:** The network of membranes that extends throughout the cell and is involved in protein synthesis and lipid metabolism.
>
> **Enzyme:** Any of numerous complex proteins that are produced by living cells and spark specific biochemical reactions.
>
> **Eukaryote:** A cell that contains a distinct nucleus and organelles.
>
> **Flagellum:** A whiplike structure that provides for movement in some cells.
>
> **Golgi body:** Organelle that sorts, modifies, and packages molecules.
>
> **Membrane:** A thin, flexible layer of plant or animal tissue that covers, lines, separates or holds together, or connects parts of an organism.
>
> **Mitochondrion:** The power-house of the cell that contains the enzymes necessary for turning food into energy.

cells.) Ribosomes are not bounded by membranes and are not considered, therefore, to be organelles.

The structure of prokaryotes. An example of a typical prokaryote is the bacterial cell. Bacterial cells can be shaped like rods, spheres, or corkscrews. Like all cells, prokaryotes are bounded by a plasma membrane. Surrounding this plasma membrane is a cell wall. In addition, in some bacteria, a jelly-like material known as a capsule coats the cell wall. Many disease-causing bacteria have capsules. The capsule provides an extra layer of protection for the bacteria. Pathogenic bacteria with capsules tend to cause much more severe disease than those without capsules.

Nuclear envelope: The double membrane that surrounds the nucleus.

Nuclear pore: Tiny openings that stud the nuclear envelope.

Nucleolus: The darker region within the nucleolus where ribosomal subunits are manufactured.

Nucleus: The control center of a cell that contains the DNA.

Organelle: A membrane-bounded cellular "organ" that performs a specific set of functions within a eukaryotic cell.

Pili: Short projections that assist bacteria in attaching to tissues.

Plasma membrane: The membrane of a cell.

Plastid: A vesicle-like organelle found in plant cells.

Prokaryote: A cell without a true nucleus.

Protein: Large molecules that are essential to the structure and functioning of all living cells.

Protist: A single-celled eukaryotic organism.

Ribosome: A protein composed of two subunits that functions in protein synthesis.

Vacuole: A space-filling organelle of plant cells.

Vesicle: A membrane-bound sphere that contains a variety of substances in cells.

Within the cytoplasm of prokaryotes is a nucleoid, a region where the cell's genetic material is stored. (Genes determine the characteristics passed on from one generation to the next.) The nucleoid is not a true nucleus because it is not surrounded by a membrane. Also within the cytoplasm are numerous ribosomes.

Attached to the cell wall of some bacteria are flagella, whiplike structures that make it possible for the bacteria to move. Some bacteria also have pili, short, fingerlike projections that help the bacteria to attach to tissues. Bacteria cannot cause disease if they cannot attach to tissues. Bacteria that cause pneumonia, for instance, attach to the tissues of the lung. Bacterial pili greatly facilitate this attachment to tissues. Thus,

Cell

Some features common to animal cells. (Reproduced by permission of Photo Researchers, Inc.)

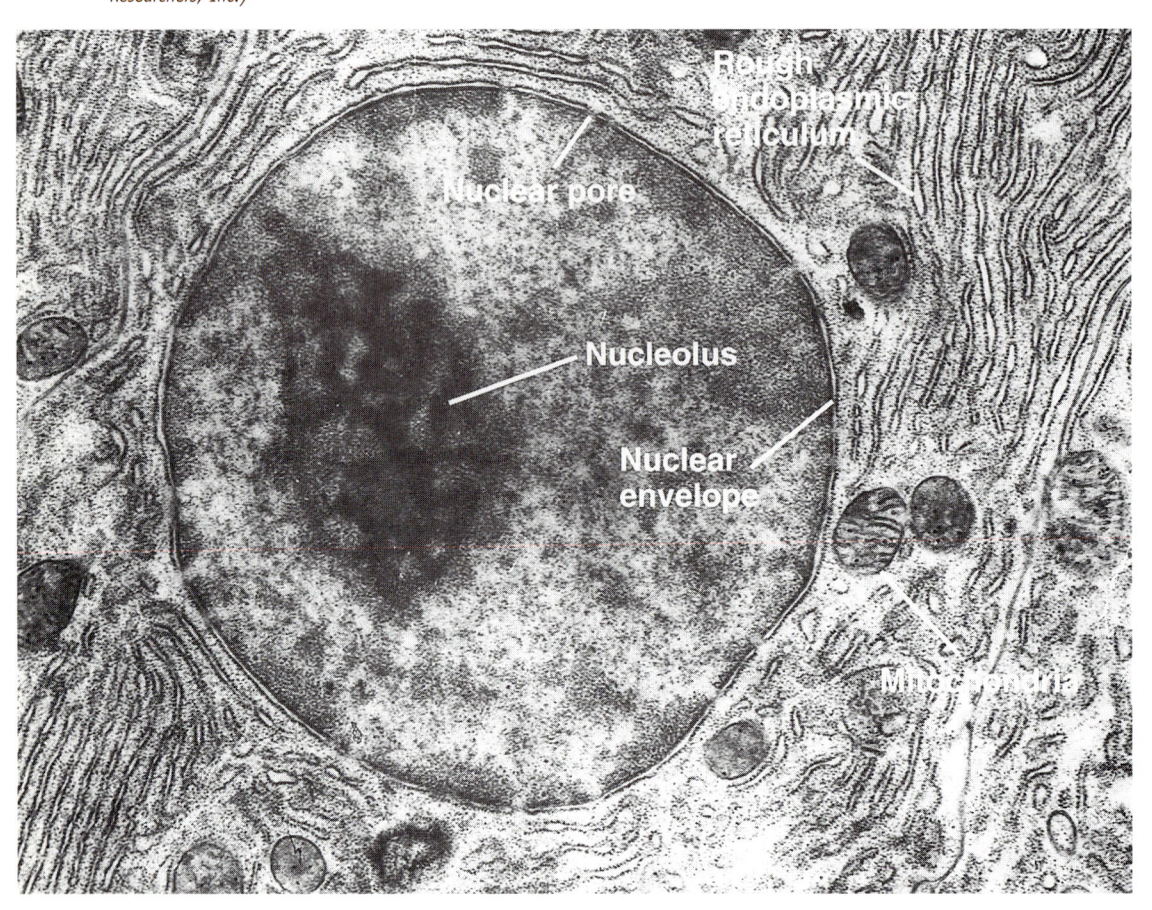

bacteria with pili, like those with capsules, are often more deadly than those without.

The structure of eukaryotes. The organelles found in eukaryotes include the membrane system, consisting of the plasma membrane, endoplasmic reticulum, Golgi body, and vesicles; the nucleus; cytoskeleton; and mitochondria. In addition, plant cells have special organelles not found in animals cells. These organelles are the chloroplasts, cell wall, and vacuoles. (See the drawing of a plant cell on page 435.)

Plasma membrane. The plasma membrane of the cell is often described as selectively permeable. That term means that some substances are able to pass through the membrane but others are not. For example, the products formed by the breakdown of foods are allowed to pass into a cell, and the waste products formed within the cell are allowed to pass out of the cell. Since the 1960s, scientists have learned a great deal about the way the plasma membrane works. It appears that some materials are able to pass through tiny holes in the membrane of their own accord. Others are helped to pass through the membrane by molecules located on the surface of and within the membrane itself. The study of the structure and function of the plasma membrane is one of the most fascinating in all of cell biology.

Endoplasmic reticulum. The endoplasmic reticulum (ER) consists of flattened sheets, sacs, and tubes of membrane that cover the entire expanse of a eukaryotic cell's cytoplasm. The ER looks something like a very complex subway or highway system. That analogy is not a bad one, since a major function of ER is to transport materials throughout the cell.

Two kinds of ER can be identified in a cell. One type is called rough ER and the other is called smooth ER. The difference between the two is that rough ER contains ribosomes on its outside surface, giving it a rough or grainy appearance. Rough ER is involved in the process of protein synthesis (production) and transport. Proteins made on the ribosomes attached to rough ER are modified, "packaged," and then shipped to various parts of the cell for use. Some are sent to the plasma membrane, where they are moved out of the cell and into other parts of the organism's body for use.

Smooth ER has many different functions, including the manufacture of lipids (fatlike materials), the transport of proteins, and the transmission of nerve messages.

The Golgi body. The Golgi body is named for its discoverer, the nineteenth century Italian scientist Camillo Golgi (1843–1926). It is one of the most unusually shaped organelles. Looking somewhat like a stack of pancakes, the Golgi body consists of a pile of membrane-bounded, flattened sacs. Surrounding the Golgi body are numerous small membrane-bounded vesicles (particles). The function of the Golgi body and its vesicles is to sort, modify, and package large molecules that are secreted by the cell or used within the cell for various functions.

The Golgi body can be compared to the shipping and receiving department of a large company. Each Golgi body within a cell has a cis face, which is similar to the receiving division of the department. Here, the Golgi body receives molecules manufactured in the endoplasmic reticulum. The trans face of the Golgi body can be compared to the shipping division of the department. It is the site from which modified and packaged molecules are transported to their destinations.

Vesicles. Vesicles are small, spherical particles that contain various kinds of molecules. Some vesicles, as noted above, are used to transport molecules from the endoplasmic reticulum to the Golgi body and from the Golgi body to various destinations. Special kinds of vesicles perform other functions as well. Lysosomes are vesicles that contain enzymes involved in cellular digestion. Some protists, for instance, engulf other cells for food. In a process called phagocytosis (pronounced FA-go-sy-to-sis), the protist surrounds a food particle and engulfs it within a vesicle. This food-containing vesicle is transported within the protist's cytoplasm until it is brought into contact with a lysosome. The food vesicle and lysosome merge, and the enzymes within the lysosome are released into the food vesicle. The enzymes break down the food into smaller parts for use by the protist.

The nucleus. The nucleus is the control center of the cell. Under a microscope, the nucleus looks like a dark blob, with a darker region, called the nucleolus, centered within it. The nucleolus is the site where parts of ribosomes are manufactured. Surrounding the nucleus is a double membrane called the nuclear envelope. The nuclear envelope is covered with tiny openings called nuclear pores.

The nucleus directs all cellular activities by controlling the synthesis of proteins. Proteins are critical chemical compounds that control almost everything that cells do. In addition, they make up the material from which cells and cell parts themselves are made.

The instructions for making proteins are stored inside the nucleus in a helical molecule called deoxyribonucleic acid, or DNA. DNA molecules differ from each other on the basis of certain chemical units, called nitrogen bases, that they contain. The way nitrogen bases are arranged within any given DNA molecule carries a specific genetic "message." One arrangement of nitrogen bases might carry the instruction "Make protein A," another arrangement of bases might carry the message "Make protein B," yet a third arrangement might code for the message "Make protein C," and so on.

The first step in protein synthesis begins in the nucleus. Within the nucleus, DNA is translated into a molecule called messenger ribonucleic acid (mRNA). MRNA then leaves the nucleus through the nuclear pores. Once in the cytoplasm, mRNA attaches to ribosomes and initiates protein synthesis. The proteins made on ribosomes may be used within the same cell or shipped out of the cell through the plasma membrane for use by other cells.

The cytoskeleton. The cytoskeleton is the skeletal framework of the cell. Instead of bone, however, the cell's skeleton consists of three kinds of protein filaments that form networks. These networks give the cell shape and provide for cellular movement. The three types of cytoskeletal fibers are microtubules, actin filaments, and intermediate filaments.

Microtubules are very thin, long tubes that form a network of "tracks" over which various organelles move within the cell. Microtubules also form small, paired structures called centrioles within animal cells. These structures are not considered organelles because they are not bounded by membranes. Centrioles are involved in the process of cell division (reproduction).

Some eukaryotic cells move about by means of microtubules attached to the exterior of the plasma membrane. These microtubules are called flagella and cilia. Cells with cilia also perform important functions in the human body. The airways of humans and other animals are lined with such cells that sweep debris and bacteria upwards, out of the lungs and into the throat. There, the debris is either coughed from the throat or swallowed into the digestive tract, where digestive enzymes destroy harmful bacteria.

Actin filaments are especially prominent in muscle cells, where they provide for the contraction of muscle tissue. Intermediate filaments are relatively strong and are often used to anchor organelles in place within the cytoplasm.

A plant cell. *(Reproduced by permission of The Gale Group.)*

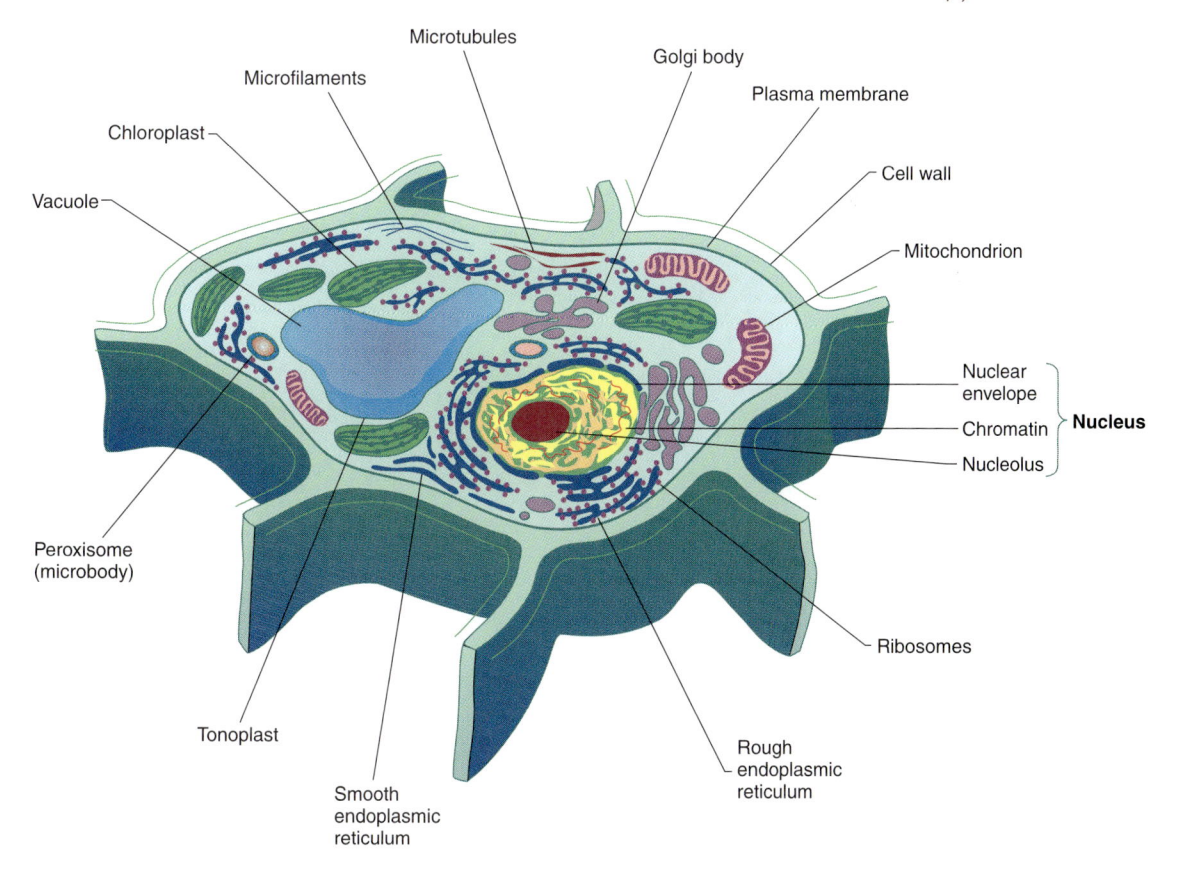

Mitochondria. The mitochondria are the power plants of cells. Each sausage-shaped mitochondrion is covered by an outer membrane. The inner membrane of a mitochondrion is folded into compartments called cristae (meaning "box"). The matrix, or inner space created by the cristae, contains the enzymes necessary for the many chemical reactions that eventually transform food molecules into energy.

Plant organelles. Plant cells have several organelles not found in animal cells. These include plastids, vacuoles, and a cell wall.

Plastids are vesicle-type organelles that perform a variety of functions in plants. For example, amyloplasts store starch and chromoplasts store pigment molecules that give some plants their vibrant orange and yellow colors. Chloroplasts are plastids that carry out photosynthesis, a process in which water and carbon dioxide are transformed into sugars.

Vacuoles are large vesicles bound by a single membrane. In many plant cells, they occupy about 90 percent of the cellular space. They perform a variety of functions in the cell, including storage of organic compounds, waste products, pigments, and poisonous compounds as well as digestive functions.

All plant cells have a cell wall that surrounds the plasma membrane. The cell wall of plants consists of a tough carbohydrate substance called cellulose laid down in a medium or network of other carbohydrates. (A carbohydrate is a compound consisting of carbon, hydrogen, and oxygen found in plants and used as a food by humans and other animals.) The cell wall provides an additional layer of protection between the contents of the cell and the outside environment. The crunchiness of an apple, for instance, is attributed to the presence of these cell walls.

[*See also* **Chromosome; Enzyme; Neuron; Nucleic acid; Protein; Reproduction; Respiration**]

Cell, electrochemical

Electrochemical cells are devices for turning chemical energy into electrical energy or, alternatively, changing electrical energy into chemical energy. The first type of cell is known as a voltaic, or galvanic, cell, while the second type is an electrolytic cell. The voltaic cell with which you are probably most familiar is the battery. Batteries consist of one or more cells connected to each other. Electrolytic cells are less common in everyday life, although they are important in many industrial operations, as in the electroplating of metals.

> ## Words to Know
>
> **Anode:** The electrode in an electrochemical cell at which electrons are given up to a reaction.
>
> **Cathode:** The electrode in an electrochemical cell at which electrons are taken up from a reaction.
>
> **Electrode:** A material that will conduct an electrical current, usually a metal, used to carry electrons into or out of an electrochemical cell.
>
> **Electrolysis:** The process by which an electrical current causes a chemical change, usually the breakdown of some substance.

History

Cells that obtain electrical energy from chemical reactions were discovered more than two centuries ago. Italian anatomist Luigi Galvani (1737–1798) first observed this effect in 1771. He noticed that the muscles of a dead frog twitched when the frog was being dissected. Galvani thought the twitching was the result of "animal electricity" that remained in the frog. Although his explanation was incorrect, credit for his observation of the effect is acknowledged in the name galvanic cell, which is sometimes used for devices of this kind.

The correct explanation for the twitching of dead frog muscles was provided by Italian physicist Alessandro Volta (1745–1827) two decades later. Volta was able to prove that the twitching was caused by an electric current that was produced when two different metals touched the animal's bloodstream at the same time. Because of Volta's contribution to this field of science, electricity-generating electrochemical cells are also called voltaic cells.

Voltaic cells

Voltaic cells contain three main components: two different metals, a solution into which the two metals are immersed, and an external circuit (such as a wire) that connects the two metals to each other.

When a metal is immersed in a solution, such as a water solution of sulfuric acid, the metal tends to lose electrons. Each metal has a greater or lesser tendency to lose electrons compared to other metals. For example, imagine that a strip of copper metal and a strip of zinc metal are both immersed in a solution of sulfuric acid. In this case, the zinc metal has a greater tendency to lose electrons than does the copper metal.

Nothing happens if you immerse two separate metals in a solution because any electrons lost by either metal have no place to go. But by attaching a wire across the top of the two metal strips, electrons are able to travel from the metal that loses them most easily (zinc in this case) across the wire to the metal that loses them less easily (copper in this case). You can observe this effect if you connect an electrical meter to the wire joining the two metals. When the metals are immersed into the sulfuric acid solution, the needle on the electrical meter jumps, indicating that an electrical current is flowing from one metal to the other.

Instead of connecting an electrical meter to the wire, you could insert an appliance that operates on electricity. For example, if a lightbulb is attached to the wire joining the two metal strips, it begins to glow. Electrons produced at the zinc strip travel through the wire and the lightbulb, causing it to light up.

Various factors determine the amount of electric current produced by a voltaic cell. The most important of these is the choice of metals used in the cell. Two metals with nearly equal tendencies to lose electrons will produce only a small current. Two metals with very different tendencies to lose electrons will produce a much larger current. Chemists have invented a measure of the tendency of various substances to lose electrons in a voltaic cell. They call that tendency the standard electrode potential for the substance. A metal with one of the highest standard electrode potentials is potassium metal, whose standard electrode potential is 2.92 volts. In comparison, a metal with a very low standard electrode potential is iron, with a value of 0.04 volt.

Electrolytic cell

An electrolytic cell is just the reverse of a voltaic cell. Rather than producing electricity by means of chemical reactions, an electrolytic cell uses electrical energy to make chemical reactions happen.

An electrolytic cell also consists of two metals immersed in a solution connected by means of an external wire. In this case, however, the external wire is hooked to some source of electrical energy, such as a battery. Electrons flow out of the battery through the wire and into one of the two metals. These two metals are known as electrodes. The electrode on which electrons accumulate is the cathode, and the electrode from which electrons are removed is the anode.

The solution in an electrolytic cell is one that can be broken apart by means of an electric current. A common example is the electrolysis of water. When an electric current flows into water, it causes water molecules to break apart, forming atoms of hydrogen and oxygen:

$$2\ H_2O \rightarrow 2\ H_2 + O_2$$

Hydrogen gas is given off at one electrode and oxygen gas at the other electrode. The electrolysis of water is, in fact, one method for making hydrogen and oxygen gas for commercial and industrial applications. Another common use for electrolytic cells is in electroplating, in which one metal is deposited on the surface of a second metal.

[*See also* **Battery; Cathode**]

Cellular/digital technology

Cellular technology is the use of wireless communication, most commonly associated with the mobile phone. The term cellular comes from the design of the system, which carries mobile phone calls from geographical service areas that are divided into smaller pockets, called cells. Each cell contains a base station that accepts and transfers the calls from mobile phones that are based in its cell. The cells are interconnected by a central controller, called the mobile telecommunications switching office (MTSO). The MTSO connects the cellular system to the conventional telephone network and it also records call information so users can be charged appropriately. In addition, the MTSO system enables the signal strength to be examined every few seconds—automatically by computer—and then be switched to a stronger cell if necessary. The user does not notice the "handoff" from one cell to another.

Traditional cellular technology uses analog service. This type of service transmits calls in one continuous stream of information between the mobile phone and the base station on the same frequency. Analog technology modulates (varies) radio signals so they can carry information such as the human voice. The major drawback to using analog service is the limitation on the number of channels that can be used.

Digital technology, on the other hand, uses a simple binary code to represent any signal as a sequence of ones and zeros. The smallest unit of information for the digital transmission system is called a bit, which

Cellular/digital technology

> **Words to Know**
>
> **Analog:** A method in which one type of data is represented by another varying physical quantity.
>
> **Cell:** Broadcasting zone in a geographical service area containing a base station that accepts and transfers calls from mobile phones that are based in the cell.
>
> **Digital:** The opposite of analog, it is a way of showing the quantity of something directly as digits or numbers.

is either one or zero. Digital technology encodes the user's voice into a bit stream. By breaking down the information into these small units, digital technology allows faster transmission of data and in a more secure form than analog.

Development of cellular/digital technology

Prior to the late 1940s, when the mobile phone was in its early stages of development, it worked like a two-way radio, using one frequency to send a user's voice back and forth. At that time, the Federal Communications Commission (FCC) limited the number of radio frequencies available for transmitting signals from a mobile phone to one transmission antenna per city. (The FCC is the government department in charge of regulating anything that goes out over the airwaves, such as telephone, television, and radio.) Approximately two dozen calls only could be placed in each city at a single time. Mobile phone users often had to wait up to a half hour to place a call. Because of the limited number of available frequencies, which allowed so few people to use the technology at one time, communication companies did not want to devote resources to researching a system that would not be widely used. They did not want to spend development money on a project that had such great limits.

During this time, communication researchers had begun to develop the cellular phone system. They theorized that by establishing service areas divided in cells and then splitting the cells into smaller geographical areas and implementing frequency reuse (where the same channel may be used for communication in cells located far apart enough to keep interference low), base stations could handle increased calls. This increase would then improve the useability of the mobile phone for consumers. However, following through on that theory was not yet possible, as the technology was still in its earliest stages.

Almost twenty years later, in 1968, the FCC reconsidered its earlier position on radio frequency limitations. In an attempt to encourage communication companies to develop better cellular systems, the FCC decided to increase the number of available radio frequencies.

Mobile phone usage takes off

In 1977, two communication companies, AT&T and Bell Labs, began the experimental use of a cellular system. By 1979, the first commercial cellular system debuted in Tokyo, Japan. In 1982, the FCC realized the incredible potential cellular communication had and authorized commercial cellular service in the United States. The first commercial analog cellular service, Advanced Mobile Phone Service (AMPS), was made available to the public in 1983 by the communication company Ameritech. Five years later, the popularity of the mobile phone skyrocketed and the number of users went beyond one million. The need to improve the constantly busy radio frequencies, however, became all too apparent for the communication companies.

Fortunately, in 1987, the FCC announced that communication companies could develop new technologies as an alternative to the then-current standard of AMPS. The FCC sought an improvement in and enlargement of cellular service. By 1991, the Telecommunications Industry Association (TIA) responded by creating personal communication services (PCS) technology. This new technology generally employed the use of all-digital wireless communication. The FCC announced in 1994 that it would allocate a spectrum (range of frequencies of sound waves) specifically for the PCS technology, which helped push the speed of the development of digital service.

Importance of cellular/digital technology

The development of cellular/digital technology is playing an increasing importance in today's business markets. More employees are spending additional time away from their offices, thereby increasing the necessity of mobile communication technologies such as the hand-held phone, notebook computers, pagers, personal digital assistants, and palm-top computers.

[*See also* **Telephone**]

Cellulose

Cellulose is the substance that makes up most of a plant's cell walls. Since it is made by all plants, it is probably the most abundant organic compound on Earth. Aside from being the primary building material for plants, cellulose has many others uses. According to how it is treated, cellulose can be used to make paper, film, explosives, and plastics, in addition to having many other industrial uses. The paper in this book contains cellulose, as do some of the clothes you are wearing. For humans, cellulose is also a major source of needed fiber in our diet.

The structure of cellulose

Cellulose is usually described by chemists and biologists as a complex carbohydrate (pronounced car-bow-HI-drayt). Carbohydrates are organic compounds made up of carbon, hydrogen, and oxygen that function as sources of energy for living things. Plants are able to make their own carbohydrates that they use for energy and to build their cell walls. According to how many atoms they have, there are several different types of carbohydrates, but the simplest and most common in a plant is glucose. Plants make glucose (formed by photosynthesis) to use for energy or to store as starch for later use. A plant uses glucose to make cellulose when it links many simple units of glucose together to form long chains. These long chains are called polysaccharides (meaning "many sugars"

Scanning electron micrograph of wood cellulose. *(Reproduced by permission of Phototake.)*

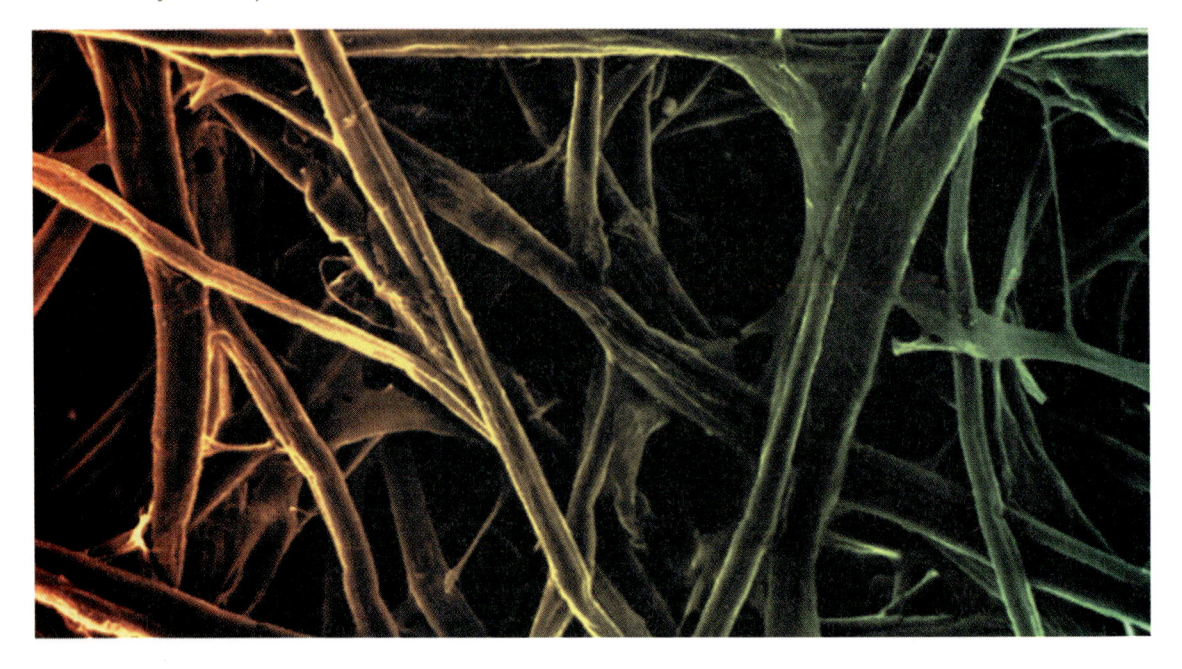

> **Words to Know**
>
> **Carbohydrate:** A compound consisting of carbon, hydrogen, and oxygen found in plants and used as a food by humans and other animals.
>
> **Glucose:** Also known as blood sugar; a simple sugar broken down in cells to produce energy.
>
> **Photosynthesis:** Chemical process by which plants containing chlorophyll use sunlight to manufacture their own food by converting carbon dioxide and water to carbohydrates, releasing oxygen as a by-product.

and pronounced pahl-lee-SAK-uh-rydes), and they form very long molecules that plants use to build their walls.

It is because of these long molecules that cellulose is insoluble or does not dissolve easily in water. These long molecules also are formed into a criss-cross mesh that gives strength and shape to the cell wall. Thus while some of the food that a plant makes when it converts light energy into chemical energy (photosynthesis) is used as fuel and some is stored, the rest is turned into cellulose that serves as the main building material for a plant. Cellulose is ideal as a structural material since its fibers give strength and toughness to a plant's leaves, roots, and stems.

Cellulose and plant cells

Since cellulose is the main building material out of which plants are made, and plants are the primary or first link in what is known as the food chain (which describes the feeding relationships of all living things), cellulose is a very important substance. It was first isolated in 1834 by the French chemist Anselme Payen (1795–1871), who earlier had isolated the first enzyme. While studying different types of wood, Payen obtained a substance that he knew was not starch (glucose or sugar in its stored form), but which still could be broken down into its basic units of glucose just as starch can. He named this new substance "cellulose" because he had obtained it from the cell walls of plants.

As the chief constituent (or main ingredient) of the cell walls of plants, cellulose performs a structural or skeletal function. Just as our hard, bony skeletons provide attachment points for our muscles and support our bodies, so the rigidity or stiffness found in any plant is due to the strength of its cell walls. Examined under a powerful microscope, fibers of cellulose are seen to have a meshed or criss-cross pattern that looks as if it were woven much as cloth. The cell wall has been likened to the way reinforced concrete is made, with the cellulose fibers acting as the rebars or steel rods do in concrete (providing extra strength). As the new cell grows, layer upon layer of new material is deposited inside the last layer, meaning that the oldest material is always on the outside of the plant.

Human uses of cellulose

Cellulose is one of the most widely used natural substances and has become one of the most important commercial raw materials. The major sources of cellulose are plant fibers (cotton, hemp, flax, and jute are almost all cellulose) and, of course, wood (about 42 percent cellulose). Since cellulose is insoluble in water, it is easily separated from the other constituents of a plant. Cellulose has been used to make paper since the Chinese first invented the process around A.D. 100. Cellulose is separated from wood by a pulping process that grinds woodchips under flowing water. The pulp that remains is then washed, bleached, and poured over a vibrating mesh. When the water finally drains from the pulp, what remains is an interlocking web of fibers that, when dried, pressed, and smoothed, becomes a sheet of paper.

Raw cotton is 91 percent cellulose, and its fiber cells are found on the surface of the cotton seed. There are thousands of fibers on each seed, and as the cotton pod ripens and bursts open, these fiber cells die. Because these fiber cells are primarily cellulose, they can be twisted to form thread or yarn that is then woven to make cloth. Since cellulose reacts easily to both strong bases and acids, a chemical process is often used to make other products. For example, the fabric known as rayon and the transparent sheet of film called cellophane are made using a many-step process that involves an acid bath. In mixtures if nitric and sulfuric acids, cellulose can form what is called guncotton or cellulose nitrates that are used for explosives. However, when mixed with camphor, cellulose produces a plastic known as celluloid, which was used for early motion-picture film. However, because it was highly flammable (meaning it could easily catch fire), it was eventually replaced by newer and more stable plastic materials. Although cellulose is still an important natural resource, many of the products that were made from it are being produced easier and cheaper using other materials.

Importance to human diet

Despite the fact that humans (and many other animals) cannot digest cellulose (meaning that their digestive systems cannot break it down into its basic constituents), cellulose is nonetheless a very important part of the healthy human diet. This is because it forms a major part of the dietary fiber that we know is important for proper digestion. Since we cannot break cellulose down and it passes through our systems basically unchanged, it acts as what we call bulk or roughage that helps the movements of our intestines. Among mammals, only those that are ruminants (cud-chewing animals like cows and horses) can process cellulose. This is because they have special bacteria and microorganisms in their digestive tracts that do it for them. They are then able to absorb the broken-down cellulose and use its sugar as a food source. Fungi are also able to break down cellulose into sugar that they can absorb, and they play a major role in the decomposition (rotting) of wood and other plant material.

[See also **Plant**]

Centrifuge

A centrifuge is a device that uses centrifugal force to separate two or more substances of different density or mass from each other. Centrifugal force is the tendency of an object traveling around a central point to fly away from that point in a straight line. A centrifuge is able to separate different substances from each other because materials with heavier masses move faster and farther away from the central point than materials with lighter masses. The first successful centrifuge was invented in 1883 by Swiss engineer Carl de Laval.

A centrifuge consists of a fixed base and center stem to which arms or holders containing hollow tubes are attached. When the device is turned on, the arms spin around the center stem at a high rate of speed. In the process, the heavier material is thrown outward within the tube while the lighter material stays near the center of the device.

Applications of the centrifuge

Large-scale centrifugation has found a great variety of commercial and industrial uses. For example, the separation of cream from whole milk has been accomplished by this process for more than a century. Today, the food, chemical, and mineral industries use centrifuges to separate water from all sorts of solids. Medical laboratories use centrifuges to separate plasma from heavier blood cells.

Modern centrifuges can even separate mixtures of different sized molecules or microscopic particles such as parts of cells. These instruments,

Centrifuge

called ultracentrifuges, spin so fast that the centrifugal force created can be more than one-half million times greater than the force of gravity.

Centrifuge studies have been very important in the development of manned space flight programs. Human volunteers are placed into very large centrifuges and then spun at high speeds. Inside the centrifuge, humans feel intense gravitational forces (g forces) similar to those that occur during the launch of space vehicles. Such experiments help space scientists understand the limits of acceleration that humans can endure in such situations.

[*See also* **Gravity and gravitation**]

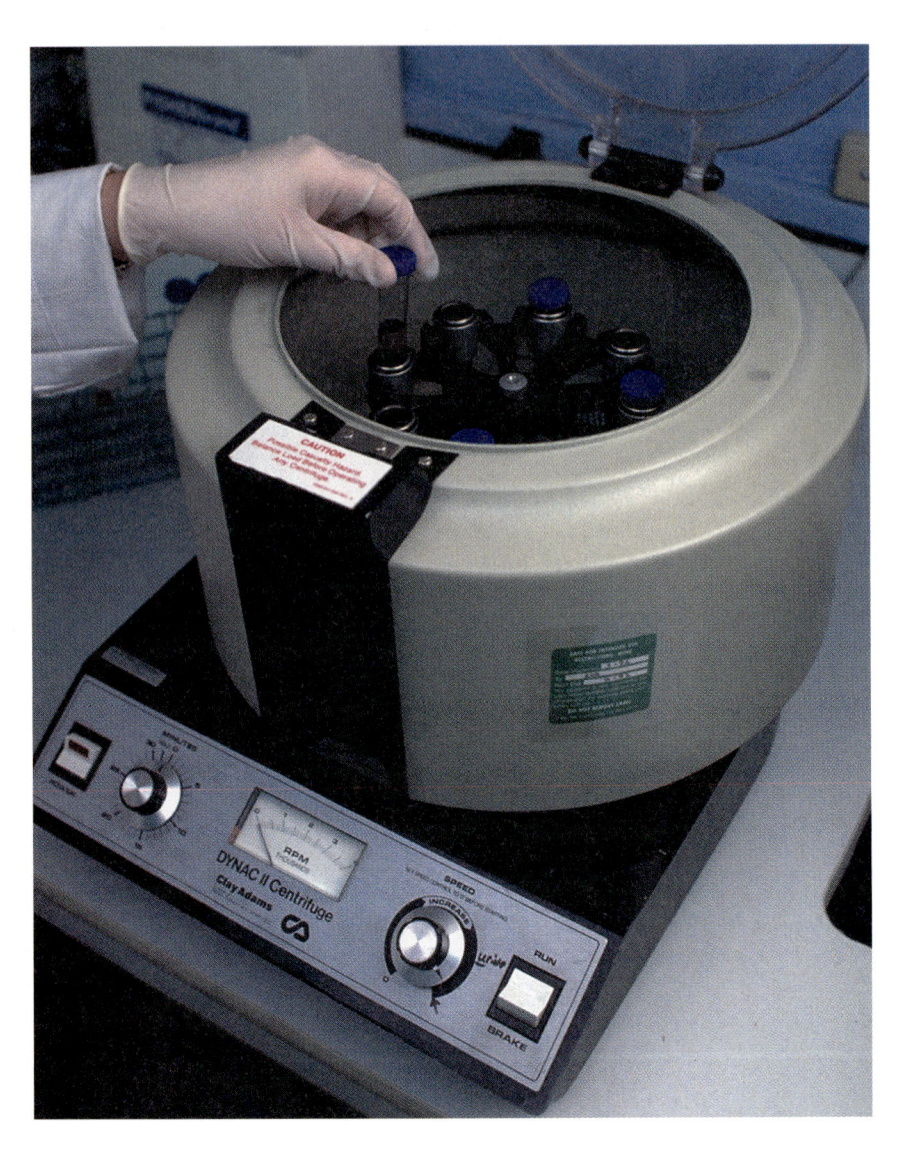

A centrifuge. *(Reproduced by permission of Photo Researchers, Inc.)*

Ceramic

Ceramic is a hard, brittle substance that resists heat and corrosion and is made by heating a nonmetallic mineral or clay at an extremely high temperature. The word ceramic comes from the Greek word for burnt material, *keramos*. Ceramics are used to produce pottery, porcelain, china, and ceramic tile. They may also be found in cement, glass, plumbing and construction materials, and spacecraft components.

The basic ingredient in all forms of ceramics are silicates, the main rock-forming minerals. Most silicates are composed of at least one type of metal combined with silicon and oxygen. Feldspar and silica are example of silicates. When silicates are combined with a liquid such as water, they form a mixture that can be kneaded and shaped into any form. After shaping, the object is dried and fired in a high-temperature oven called a kiln. A glaze (a glasslike substance that makes a surface glossy and watertight) may be added between drying and firing. From ancient days to the present, this process has remained virtually the same, except for the addition of mechanical aids.

Pottery

The oldest examples of pottery, found in Moravia (a region of the Czech Republic) and dating back to 25,000 B.C., are animal shapes made of fired clay. Potter's wheels and kilns first appeared in Mesopotamia (an ancient region in southwest Asia) around 3000 B.C. Some of the most fascinating pottery in history was made by the ancient Greeks, whose vases were skillfully decorated in the methods of black figure (black paint applied to red clay) or red figure (black paint covering all but the design, which stood out in red clay). Early Islamic potters of the Middle East produced colorful, imaginatively glazed tiles and other items. Their elaborate pictorial designs have provided archaeologists with many clues to their daily lives.

Perhaps the most renowned potters of all time are the Chinese, who developed the finest form of pottery—porcelain. Made of kaolin (pronounced KA-uh-lin; a white clay free of impurities) and petuntse (a feldspar mineral that forms a glassy cement), porcelain is fired at extremely high temperatures. The result is a high-quality material that is uniformly translucent, glasslike, and white. Porcelain was first made in China during the T'ang Dynasty (618–906).

Modern ceramics

In the twentieth century, scientists and engineers acquired a much better understanding of ceramics and their properties. During World War II

Cetaceans

(1939–45), a high demand for military materials hastened the evolution of the science of ceramics. These materials are now found in a wide variety of products, including abrasives, bathroom fixtures, and electrical insulation.

During the 1960s and 1970s, the growing fields of atomic energy, electronics, communication, and space travel increased demand for more sophisticated ceramic products. Because ceramics can withstand extreme temperatures, they have been used in gas turbines and jet engines. The undersides of the space shuttles are lined by some 20,000 individually contoured silica fiber tiles that are bonded to a felt pad. The felt pad in turn is bonded to the body of a shuttle. These ceramic tiles can withstand a maximum surface temperature of 1,200 to 1,300°F (650 to 705°C).

In 1990, a team of Japanese scientists working for their government developed a stretchable material from silicon-based compounds. When made into strips and heated, this special ceramic material can be stretched to two-and-a-half times its original length without losing its hardness and durability.

Cetaceans

Cetaceans (pronounced sih-TAY-shuns) include whales, dolphins, and porpoises. Although ancient people believed they were fish, cetaceans are aquatic mammals that bear live young, produce milk to feed their offspring, and have a bit of hair. The study of cetaceans is called cetology.

Although they have a fishlike shape, cetaceans are descended from land animals and still retain some modified features of their ancestors. They have the remains of a pelvic girdle, and the bones beneath their forelimbs, which are now used as flippers for swimming, show that they once had five fingers. Like land animals they have lungs, but instead of nostrils cetaceans breathe air through blowholes on the top of their head. They have no hind legs, and their tail has developed over time into a horizontal fluke (a flat tail), used to propel the animal through the water. Other physical changes include the addition of a thick layer of blubber to insulate against the cold of the ocean depths.

Cetaceans belong to the order Cetacea and include the baleen whales (ten species that live in the ocean) and the toothed whales, whose many species (including dolphins and porpoises) are found in diverse habitats from deep ocean to freshwater rivers.

Baleen whales

Baleen (pronounced buh-LEEN) whales are huge creatures that include the blue whale, the largest animal that has ever lived and that can

> **Words to Know**
>
> **Baleen:** A flexible, horny substance making up two rows of plates that hang from the upper jaws of baleen whales.
>
> **Echolocation:** A method of locating objects by the echoes reflected from sounds produced by certain animals.
>
> **Whaling:** The harvesting of whales for their products by individuals or commercial operations.

reach a length of 100 feet (30 meters). Together with the finback, sei, humpback, minke, and gray whales, the blue whale belongs to a family of whales that migrates from cold polar waters to breed in warmer waters. The right whale, so-named because it was the "right one" for whalers, does not have the dorsal fin of the other great whales. Baleen whales are toothless; they eat by filtering tiny sea animals, or plankton, through rows of flexible, horny plates called baleen that hang from their upper jaw. One mouthful of water may contain millions of tiny prey. Baleen whales have two blowholes (toothed whales have only one), and some species have deep grooves on their throat and belly.

Toothed whales

The faster-moving, smaller-bodied toothed whales pursue squid, fish, and, in the case of killer whales, sea birds and other mammals. Killer whales have even been observed ganging up on and killing the much bigger gray whale. The largest of the toothed whales is the sperm whale. It is also the deepest diver, plunging to depths of over half a mile in search of giant squid. Its large, square head contains a cavity of oil that was used in oil lamps before petroleum became available. Certain species of toothed whales are very social, traveling in groups of dozens of animals. Bottlenosed dolphins and killer whales typically have social bonds with many others of their species that may last for life.

Sensory perception

Cetaceans have good vision and excellent hearing. Toothed whales, porpoises, and dolphins navigate and find food using echolocation, in

Cetaceans

which they produce pulses of sound and listen for the echo. By deciphering objects based on the reflected echoes, these animals can obtain an accurate picture of their physical environment.

Intelligence and communication

Cetaceans have relatively large brains and are highly intelligent animals that exhibit curiosity, affection, jealousy, self-control, sympathy, spite, and trick-playing. They communicate with each other by producing a great variety of sounds, from the moans and knocks of gray whales to the eerie songs of humpbacks.

Commercial whaling and other threats

Commercial whaling has had a devastating effect on the world's great whale populations. Modern whaling methods may be viewed as incredibly inhumane. Whales today are killed by a harpoon that has a head with four claws and one or more grenades attached. When a whale is within range, the harpoon is fired into its body, and the head of the harpoon explodes, tearing apart muscle and organs. The whale dives to escape, but is hauled to the surface with ropes and is shot once again. Death

A humpback whale in the Atlantic Ocean off the coast of Massachusetts. *(Reproduced by permission of AP/Wide World Photos.)*

may not be immediate, and the whale may struggle to live for 15 minutes of more. Although many nations have agreed to end or curtail their whaling practices, some have not.

Further dangers to cetacean species include (1) drift nets (now outlawed), in which they can become entangled, and (2) the use of purse-seines (pronounced sane; nets whose ends are pulled together to form a huge ball) for catching tuna, a method that has killed an estimated seven million dolphins since 1959. Marine pollution and the loss of food sources due to human activity are a continuing danger.

Chaos theory

Chaos theory is the study of complex systems that, at first glance, appear to follow no orderly laws of mathematics or science. Chaos theory is one of the most fascinating and promising developments in late-twentieth-century mathematics and science. It provides a way of making sense out of phenomena such as weather patterns that seem to be totally without organization or order.

Cause-and-effect and chaos

Scientists have traditionally had a rather strict cause-and-effect view of the natural world. English physicist Isaac Newton once said that if he could know the position and motion of every particle in the universe at any one moment, he could predict the future of the universe into the infinite future. He believed that all those particles follow strict physical laws. Since he knew (or so he thought) what those laws were, all he had to do was to apply them to the particles at any one point in time.

On the other hand, scientists have always realized that some events in nature appear to be just too complex to analyze by the laws of science. One of the best examples is weather patterns. Even though scientists know a great deal about the elements that make up weather, they have a very difficult time predicting what weather patterns will be. The term chaos has often been used to describe systems that are just too "messy" to understand by scientific analysis.

Origins of chaos theory

The rise of modern chaos theory can be traced to a few particularly striking and interesting discoveries. One of these events occurred in the

Chaos theory

> ### Words to Know
>
> **Attractor:** An element in a chaotic system that appears to be responsible for helping the system to settle down.
>
> **Cause-and-effect:** The view that humans can understand why certain events (effects) take place.
>
> **Chaos:** Some behavior that appears to be so complex as to be incapable of analysis by humans.
>
> **Chaos theory:** Mathematical and scientific efforts to provide cause-and-effect explanations for chaotic behavior.
>
> **Generator:** Elements in a system that appear to be responsible for chaotic behavior in the system.
>
> **Law:** A statement in science that summarizes how some aspect of nature is likely to behave. Laws have survived many experimental tests and are believed to be highly dependable.

1890s when French mathematician Henri Poincaré was working on the problem of the interactions of three planets with one another. The problem should have been fairly straightforward, Poincaré thought, since the gravitational laws involved were well known. The results of his calculations were so unexpected, however, that he gave up his work. He described those results as "so bizarre that I cannot bear to contemplate them."

Dutch engineer B. van der Pol encountered a similar problem in working with electrical circuits. He started out with systems that could easily be described by well-known mathematical equations. But the circuits he actually produced gave off unexpected and irregular noises for which he could not account.

Then, in 1961, American meteorologist Edward Lorenz found yet another example of chaotic behavior. Lorenz developed a system for predicting the weather based on 12 equations. The equations represented the factors we know to affect weather patterns, including atmospheric pressure, temperature, and humidity. What Lorenz found was that by making very small changes in the initial numbers used in these equations, he could produce wildly different results.

Generators and attractors

Scientists and mathematicians now view chaotic behavior in a different way. Instead of believing that such behavior is too complex ever to understand, they have come to conclude that certain patterns exist within chaos that can be discovered and analyzed. For example, certain characteristics of a system appear to be able to generate chaotic behavior. Such characteristics are known as generators because they cause the chaotic behavior. Very small differences in a generator can lead to very large differences in a system at a later point in time.

Researchers have also found that chaotic behavior sometimes has a tendency to settle down to some form of predictable behavior. When this happens, elements within the system appear to bring various aspects of the chaos together into a more understandable pattern. Those elements are given the name attractors because they appear to attract the parts of a chaotic system to themselves.

Applications

In theory, studies of chaos have a great many possible applications. After all, much of what goes on in the world around us seems more like chaos than a neat orderly expression of physical laws. The weather may be the best everyday example of that point. Although we know a great deal about all the elements of which weather patterns are made, we still have relatively modest success in predicting how those elements will come together to produce a specific weather pattern. Studies of chaos theory may improve these efforts.

Animal behavior also appears to be chaotic. Population experts would like very much to know how groups of organisms are likely to change over time. And, again, we know many of the elements that determine those changes, including food supplies, effects of disease, and crowding. Still, predictions of population changes—whether of white deer in the wilds of Vermont or the population of your hometown—tend to be quite inaccurate. Again, chaos theory may provide a way of making more sense out of such apparently random behavior.

Chemical bond

A chemical bond is any force of attraction that holds two atoms or ions together. In most cases, that force of attraction is between one or more negatively charged electrons held by one of the atoms and the positively

> ## Words to Know
>
> **Covalent bond:** A chemical bond formed when two atoms share one or more pairs of electrons with each other.
>
> **Double bond:** A covalent bond consisting of two pairs of electrons.
>
> **Electronegativity:** A numerical method for indicating the relative tendency of an atom to attract the electrons that make up a covalent bond.
>
> **Hydrogen bond:** A chemical bond formed between two atoms or ions with opposite charges.
>
> **Ionic bond:** A chemical bond formed when one atom gains and a second atom loses electrons. An ion is a molecule or atom that has lost one or more electrons and is, therefore, electrically charged.
>
> **Multiple bond:** A double or triple bond.
>
> **Polar bond:** A covalent bond in which one end of the bond is more positive than the other end.
>
> **Triple bond:** A covalent bond consisting of three pairs of electrons.

charged nucleus of the second atom. Chemical bonds vary widely in their strength, ranging from relatively strong covalent bonds (in which electrons are shared between atoms) to very weak hydrogen bonds. The term chemical bond also refers to the symbolism used to represent the force of attraction between two atoms or ions. For example, in the chemical formula H—O—H, the short dashed lines are known as chemical bonds.

History

Theories of chemical bonds go back a long time. One of the first was developed by Roman poet Lucretius (c. 95–c. 55 B.C.), author of *De Rerum Natura* (title means "on the nature of things"). In this poem, Lucretius described atoms as tiny spheres with fishhook-like arms. Atoms combined with each other, according to Lucretius, when the hooked arms of two atoms became entangled with each other.

Such theories were pure imagination, however, for many centuries, since scientists had no true understanding of an atom's structure until the beginning of the twentieth century. It was not until then that anything approaching a modern theory of chemical bonding developed.

Covalent bonding

Today, it is widely accepted that most examples of chemical bonding represent a kind of battle between two atoms for one or more electrons. Imagine an instance, for example, in which two hydrogen atoms are placed next to each other. Each atom has a positively charged nucleus and one electron spinning around its nucleus. If the atoms are close enough to each other, then the electrons of both atoms will be attracted by both nuclei. Which one wins this battle?

The answer may be obvious. Both atoms are exactly identical. Their nuclei will pull with equal strength on both electrons. The only possible result, overall, is that the two atoms will share the two electrons with each other equally. A chemical bond in which two electrons are shared between two atoms is known as a covalent bond.

Ionic bonding

Consider now a more difficult situation, one in which two different atoms compete for electrons. One example would be the case involving a sodium atom and a chlorine atom. If these two atoms come close enough to each other, both nuclei pull on all electrons of both atoms. In this case, however, a very different result occurs. The chlorine nucleus has a much larger charge than does the sodium nucleus. It can pull on sodium's electrons much more efficiently than the sodium nucleus can pull on the chlorine electrons. In this case, there is a winner in the battle: chlorine is able to pull one of sodium's electrons away. It adds that electron to its own collection of electrons. In a situation in which one atom is able to completely remove an electron from a second atom, the force of attraction between the two particles is known as an ionic bond.

Electronegativity

Most cases of chemical bonding are not nearly as clear-cut as the hydrogen and the sodium/chlorine examples given above. The reason for this is that most atoms are more nearly matched in their ability to pull electrons than are sodium and chlorine, although not as nearly matched as two identical atoms (such as two hydrogen atoms).

A method for expressing the pulling ability of two atoms was first suggested by American chemist Linus Pauling (1901–1994). Pauling proposed the name "electronegativity" for this property of atoms. Two atoms with the same or similar electronegativities will end up sharing electrons between them in a covalent bond. Two atoms with very different electronegativities will form ionic bonds.

Polar and nonpolar bonds

In fact, most chemical bonds do not fall into the pure covalent or pure ionic bond category. The major exception occurs when two atoms of the same kind—such as two hydrogen atoms—combine with each other. Since the two atoms have the same electronegativities, they must share electrons equally between them.

Consider the situation in which aluminum and nitrogen form a chemical bond. The electronegativity difference between these two atoms is about 1.5. (For comparison's sake, the electronegativity difference between sodium and chlorine is 2.1 and between hydrogen and hydrogen is 0.0.) A chemical bond formed between aluminum and nitrogen, then, is a covalent bond, but electrons are not shared equally between them. Instead, electrons that make up the bond spend more of their time with nitrogen (which pulls more strongly on electrons) than with aluminum (which pulls less strongly). A covalent bond in which electrons spend more time with one atom than with the other is called a polar covalent bond. In contrast, a bond in which electrons are shared equally (as in the case of hydrogen) is called a nonpolar covalent bond.

Multiple bonds

All covalent bonds, polar and nonpolar, always consist of two electrons. In some cases, both electrons come from one of the two atoms. In most cases, however, one electron comes from each of the two atoms joined by the bond.

In some cases, atoms may share more than two electrons. If so, however, they still share pairs only: two pairs or three pairs, for example. A bond consisting of two pairs of (that is, four) electrons is called a double bond. One containing three pairs of electrons is called a triple bond.

Other types of bonds

Other types of chemical bonds also exist. The atoms that make up a metal, for example, are held together by a metallic bond. A metallic bond is one in which all of the metal atoms share with each other a cloud of electrons. The electrons that make up that cloud originate from the outermost energy levels of the atoms.

A hydrogen bond is a weak force of attraction that exists between two atoms or ions with opposite charges. For example, the hydrogen-oxygen bonds in water are polar bonds. The hydrogen end of these bonds are slightly positive, and the oxygen ends are slightly negative. Two molecules of water placed next to each other will feel a force of attraction because the oxygen end of one molecule feels an electrical force of attraction to the hydrogen end of the other molecule. Hydrogen bonds are very common and extremely important in biological systems. They are strong enough to hold substances together but weak enough to break apart and allow chemical changes to take place within the system.

Van der Waals forces are yet another type of chemical bond. They are named in honor of the Dutch physicist Johannes Diderik van der Waals (1837–1923), who investigated the weak nonchemical bond forces between molecules. Such forces exist between particles that appear to be electrically neutral. The electrons in such particles shift back and forth very rapidly. That shifting of electrons means that some parts of the particle are momentarily charged, either positively or negatively. For this reason, very weak, short-term forces of attraction can develop between particles that are actually neutral.

Chemical warfare

Chemical warfare involves the use of natural or synthetic (human-made) substances to disable or kill an enemy or to deny them the use of resources such as agricultural products or foliage in which to hide. The effects of the chemicals may last only a short time, or they may result in permanent damage and death. Most of the chemicals used are known to be toxic (poisonous) to humans or plant life. In some cases, normally harmless chemicals have also been used to damage an enemy's environment. Such actions have been called ecocide and are one method for disrupting an enemy's economic system. The deliberate dumping of large quantities of crude oil on the land or in the ocean is an example of ecocide.

The appeal of chemicals as agents of warfare is their ability to cause mass casualties or damage to an enemy with only limited risk to the forces using the chemicals. Poisoning a town's water supply, for example, poses almost no threat to an attacking army. Yet the action could result in the death of thousands of the town's defenders. In many cases, chemicals are not detectable by the enemy until it is too late for them to take action.

History

Chemical warfare dates back to the earliest use of weapons. Poisoned arrows and darts used for hunting by primitive peoples have also been used as weapons in battles between tribal groups. In 431 B.C., the

Chemical warfare

> ### Words to Know
>
> **Defoliant:** A chemical that kills the leaves of plants and causes them to fall off.
>
> **Ecocide:** Deliberate attempts to destroy or damage the environment over a large area as a tactical element of a military strategy.
>
> **Harassing agent:** A chemical that causes temporary damage to animals, including humans.
>
> **Herbicide:** A chemical that kills entire plants, often selectively.
>
> **Nerve agent:** A chemical that kills animals, including humans, by attacking the nervous system and causing the disruption of vital functions such as respiration and heartbeat.

Spartans used burning sulfur and pitch to produce clouds of suffocating sulfur dioxide in their sieges against Athenian cities. When the Romans defeated the Carthaginians of North Africa in 146 B.C., they destroyed the city of Carthage and spread salt on surrounding fields to destroy the agricultural capability of the land. The Romans' intent was to prevent the Carthaginians from rebuilding their city.

Types of chemical agents

Chemical agents can be classified into several general categories, ranging from those that cause relatively little harm to those that can cause death. One group includes those that produce only temporary damage. As an example, tear gas tends to cause coughing, sneezing, and general respiratory discomfort, but this discomfort passes within a relatively short period of time.

Other agents cause violent skin irritation and blistering and may result in death. Still other agents are poisonous and are absorbed into the victim's bloodstream through the lungs or skin, causing death. Nerve agents attack the nervous system and kill by causing the body's vital functions (respiration, circulation, etc.) to cease. Finally, other agents cause psychological reactions including disorientation and hallucinations.

Another group of chemical agents include those that attack vegetation, damaging or killing plants. Some examples include defoliants that kill a plant's leaves, herbicides that kill the entire plant, and soil sterilants that prevent the growth of new vegetation.

Antipersonnel agents: chemicals used against people. The first large-scale use of poisonous chemicals in warfare occurred during World War I (1914–18). More than 100,000 tons (90,000 metric tons) of lethal chemicals were used by both sides in an effort to break the stalemate of endless trench warfare. The most commonly used chemicals were four lung-destroying poisons: chlorine, chloropicrin, phosgene, and trichloromethyl chloroformate, along with a skin-blistering agent known as mustard gas, or bis(2-chloroethyl) sulfide. These poisons caused about 100,000 deaths and another 1.2 million injuries, almost all of which involved military personnel.

In 1925, many of the world's nations signed an agreement, called the Geneva Protocol, to discontinue production of chemical agents for military use. Despite this agreement, the United States, Britain, Japan,

American troops wearing gas masks during World War I. The soldier at left, unable to get his mask on in time, clutches his throat as he breathes in the poisonous gas. *(Reproduced by permission of the Corbis Corporation [Bellevue].)*

Chemical warfare

Germany, Russia, and other countries all continued development of these weapons during the period between World War I and World War II (the 1920s and most of the 1930s). This research included experimentation on animals and humans. Although chemical weapons were not used very widely during World War II (1939–45), the opposing sides had large stockpiles ready to deploy against military and civilian targets.

During the civil war in Vietnam, the U.S. military used a "harassing agent" during many of its operations. (The United States sided with and supplied the South Vietnamese in the early 1960s and joined their military efforts against the North in 1964.) The agent was a tear gas known as CS or o-chlorobenzolmalononitrile. CS was not regarded as toxic to humans and was intended only to make an area uninhabitable for 15 to 45 days. A total of about 9,000 tons (8,000 metric tons) of CS were sprayed over 2.5 million acres (1.0 million hectares) of South Vietnam. Although CS was classified as nonlethal (not deadly), several hundred deaths were later reported when the gas was used in heavy concentrations in confined spaces such as underground bunkers and bomb shelters.

Poisonous chemicals were also used during the Iran-Iraq War of 1981–87, especially by Iraqi forces. During that war, both soldiers and civilians were targets of chemical weapons. Perhaps the most famous incident was the gassing of Halabja, a town in northern Iraq that had been overrun by Iranian-supported Kurds. The Iraqi military attacked Halabja with two fast-acting neurotoxins, sarin and tabun. Sarin and tabun cause rapid death by interfering with the transmission of nerve impulses. Muscular spasms develop and a person dies when he or she is no longer able to breathe. About 5,000 people, mostly civilians, were killed in this incident.

Use of herbicides during the Vietnam War. Herbicides are chemicals that were originally developed to kill weeds. However, they are just as effective at killing agricultural crops as they are at killing weeds. During the Vietnam War, in addition to tear gas, the U.S. military relied heavily on the use of herbicides as a weapon of war. The purpose of using herbicides was twofold: first, to destroy enemy crops and disrupt their food supply, and second, to remove forest cover in which enemy troops might hide. Between 1961 and 1971, about 3.2 million acres (1.3 million hectares) of forest and 247,000 acres (100,000 hectares) of Vietnamese croplands were sprayed at least once. This area is equivalent to about one-seventh of the total land area of South Vietnam.

The most commonly used herbicide was called Agent Orange, a blend of two herbicides known as 2,4-D and 2,4,5-T. Two other herbicides, picloram and cacodylic acid, were also used, but in much smaller amounts. In total, about 25,000 tons of 2,4-D, 21,000 tons of 2,4,5-T, and 1,500 tons of picloram were utilized as a result of U.S. military actions during the war.

In particular, Agent Orange was sprayed at a rate of about 22.3 pounds per acre (25 kilograms per hectare). This rate is equivalent to about 10 times the rate at which those same chemicals are used for plant control purposes in forestry. The higher spray rate was used in Vietnam because the intention of the U.S. military was the ultimate destruction of Vietnamese ecosystems (its communities of plants and animals).

The ecological damages caused by the military use of herbicides in Vietnam were not studied in detail. However, a few casual surveys have been made by some visiting ecologists. These scientists observed that coastal mangrove forests (tropical trees and shrubs that form dense greenery) were especially sensitive to treatment with herbicides. About 36 percent of the mangrove ecosystem of South Vietnam was sprayed with

Chemical warfare

Soldiers at Assaf Harofe Hospital washing "victims" in a simulated chemical warfare attack. *(Reproduced by permission of the Corbis Corporation [Bellevue].)*

Chemical warfare

herbicides, a total of about 272,000 acres (110,000 hectares). Almost all of the plant species of mangrove forests proved to be highly vulnerable to herbicides, including the dominant species of tree, red mangrove.

Severe ecological effects of herbicide spraying were also observed in the biodiverse upland forests of Vietnam, especially its rain forests. Mature tropical forests in this region have many species of hardwood trees. These forests are covered by a dense canopy consisting of complex layers. As a result, a single spraying of herbicide typically kills only about 10 percent of the larger trees. However, the goal of the U.S. military was to achieve a more extensive and longer-lasting defoliation. Hence, they sprayed many areas more than once. In fact, about 34 percent of Vietnam was treated with herbicides more than once.

The effects on animals of herbicide spraying in Vietnam are not well documented. However, there are many accounts of reduced populations of birds, mammals, reptiles, and other animals in the mangrove forests treated with herbicides. In addition, large decreases in the yield of nearshore fisheries have been attributed to the spraying of mangrove ecosystems, which provide spawning and nursery habitat for the fish.

The effects on wild animals were probably caused mostly by habitat changes resulting from herbicide spraying. However, there have also been numerous reports of domesticated agricultural animals becoming ill or dying. Because of the constraints of warfare, the specific causes of these illnesses and deaths were never studied properly by veterinary scientists. However, these ailments were commonly attributed to toxic effects of exposure to herbicides, mostly ingested by the animals with their food.

Use of petroleum as a weapon during the Persian Gulf War. Large quantities of petroleum are often spilled at sea during warfare, mostly as the result of damage to oil tankers or other facilities such as offshore production platforms. During the Iran-Iraq War of the 1980s and the Persian Gulf War of 1991–92, however, oil spills were deliberately used to gain military advantage, as well as to inflict economic damages on the enemy's postwar economy.

The world's all-time largest oceanic spill of petroleum occurred during the Persian Gulf War. The Iraqi military deliberately released almost 1 million tons (900,000 metric tons) of crude oil into the Persian Gulf from several tankers and an offshore facility for loading tankers. In part, the oil was spilled to establish a defensive barrier against an expected attack by the anti-Iraqi coalition forces. The hope was that igniting the immense quantities of spilled petroleum would create a floating inferno that would provide an effective barrier against a seaborne invasion. It is believed that the Iraqis also sought to contaminate the seawater used in desalination plants (salt removal facilities) that supply most of Saudi Arabia with freshwater.

Controls over the use of chemical weapons

The first treaty to control the use of chemical weapons was the Geneva Protocol, agreed upon in 1925 and subsequently signed by 132 nations. This treaty was prompted by the horrible uses of chemical weapons during World War I. It banned the use of asphyxiating (suffocating), poisonous, or other gases, as well as bacteriological (germ) methods of warfare. In spite of having signed this treaty, however, all major nations are known to have continued research on new and more effective chemical and bacteriological weapons.

In 1993, negotiators for various nations met at a Chemical Weapons Convention and agreed to the destruction of all chemical weapons within a 10 to 15 year period following ratification of a chemical weapons treaty. By the end of 2000, 174 nations had signed, ratified, or acceded to the treaty. In the long run, its effectiveness depends upon its ratification by all countries having significant stockpiles of chemical weapons, the countries' commitment to following the terms of the treaty, and the power of an international monitoring program to expose and discipline member countries ignoring the treaty.

Part of the problem in obtaining an effective chemical weapons treaty is desire. Nations have to want to destroy their stockpiles of weapons and discontinue making more of them. Another part of the problem is cost. By one estimate, it will cost $16 to $20 billion just to safely destroy the chemical weapons of the world's two largest military powers, the United States and Russia.

[*See also* **Agent Orange; Agrochemicals; Poisons and toxins**]

Chemistry

Chemistry is the study of the composition of matter and the changes that take place in that composition. If you place a bar of iron outside your window, the iron will soon begin to rust. If you pour vinegar on baking soda, the mixture fizzes. If you hold a sugar cube over a flame, the sugar begins to turn brown and give off steam. The goal of chemistry is to understand the composition of substances such as iron, vinegar, baking soda, and sugar and to understand what happens during the changes described here.

Chemistry

Words to Know

Analytical chemistry: That area of chemistry that develops ways to identify substances and to separate and measure the components in a compound or mixture.

Inorganic chemistry: The study of the chemistry of all the elements in the periodic table except for carbon.

Organic chemistry: The study of the chemistry of carbon compounds.

Physical chemistry: The branch of chemistry that investigates the properties of materials and relates these properties to the structure of the substance.

Qualitative analysis: The analysis of compounds and mixtures to determine the elements present in a sample.

Quantitative analysis: The analysis of compounds and mixtures to determine the percentage of elements present in a sample.

History

Both the term chemistry and the subject itself grew out of an earlier field of study known as alchemy. Alchemy has been described as a kind of pre-chemistry, in which scholars studied the nature of matter—but without the formal scientific approach that modern chemists use. The term alchemy is probably based on the Arabic name for Egypt, *al-Kimia,* or the "black country."

Ancient scholars learned a great deal about matter, usually by trial-and-error methods. For example, the Egyptians mastered many technical procedures such as making different types of metals, manufacturing colored glass, dying cloth, and extracting oils from plants. Alchemists of the Middle Ages (400–1450) discovered a number of elements and compounds and perfected other chemical techniques, such as distillation (purifying a liquid) and crystallization (solidifying substances into crystals).

The modern subject of chemistry did not appear, however, until the eighteenth century. At that point, scholars began to recognize that research on the nature of matter had to be conducted according to certain specific rules. Among these rules was one stating that ideas in chemistry had to be subjected to experimental tests. Some of the founders of modern chemistry include English natural philosopher Robert Boyle (1627–1691), who set down certain rules on chemical experimentation; Swedish chemist Jöns Jakob Berzelius (1779–1848), who devised chemical symbols, determined atomic weights, and discovered several new elements; English chemist John Dalton (1766–1844), who proposed the first modern atomic theory; and French chemist Antoine-Laurent Lavoisier (1743–1794), who first explained correctly the process of combustion (or burning), established modern terminology for chemicals, and is generally regarded as the father of modern chemistry.

Goals of chemistry

Chemists have two major goals. One is to find out the composition of matter: to learn what elements are present in a given sample and in what percentage and arrangement. This type of research is known as analysis. A second goal is to invent new substances that replicate or that are

In this 1964 photo, a photographic chemist conducts an experiment on dye formation in a traditional chemical laboratory. *(Reproduced courtesy of the Library of Congress.)*

Chemistry

different from those found in nature. This form of research is known as synthesis. In many cases, analysis leads to synthesis. That is, chemists may find that some naturally occurring substance is a good painkiller. That discovery may suggest new avenues of research that will lead to a synthetic (human-made) product similar to the natural product, but with other desirable properties (and usually lower cost). Many of the substances that chemistry has produced for human use have been developed by this process of analysis and synthesis.

Fields of chemistry

Today, the science of chemistry is often divided into four major areas: organic, inorganic, physical, and analytical chemistry. Each discipline investigates a different aspect of the properties and reactions of matter.

Organic chemistry. Organic chemistry is the study of carbon compounds. That definition sometimes puzzles beginning chemistry students because more than 100 chemical elements are known. How does it happen that one large field of chemistry is devoted to the study of only one of those elements and its compounds?

The answer to that question is that carbon is a most unusual element. It is the only element whose atoms are able to combine with each other in apparently endless combinations. Many organic compounds consist of dozens, hundreds, or even thousands of carbon atoms joined to each other in a continuous chain. Other organic compounds consist of carbon chains with other carbon chains branching off them. Still other organic compounds consist of carbon atoms arranged in rings, cages, spheres, or other geometric forms.

The scope of organic chemistry can be appreciated by knowing that more than 90 percent of all compounds known to science (more than 10 million compounds) are organic compounds.

Organic chemistry is of special interest because it deals with many of the compounds that we encounter in our everyday lives: natural and synthetic rubber, vitamins, carbohydrates, proteins, fats and oils, cloth, plastics, paper, and most of the compounds that make up all living organisms, from simple one-cell bacteria to the most complex plants and animals.

Inorganic chemistry. Inorganic chemistry is the study of the chemistry of all the elements in the periodic table except for carbon. Like their cousins in the field of organic chemistry, inorganic chemists have provided the world with countless numbers of useful products, including fertilizers, alloys, ceramics, household cleaning products, building materials, water softening and purification systems, paints and stains, computer chips and other electronic components, and beauty products.

The more than 100 elements included in the field of inorganic chemistry have a staggering variety of properties. Some are gases, others are solid, and a few are liquid. Some are so reactive that they have to be stored in special containers, while others are so inert (inactive) that they virtually never react with other elements. Some are so common they can be produced for only a few cents a pound, while others are so rare that they cost hundreds of dollars an ounce.

Because of this wide variety of elements and properties, most inorganic chemists concentrate on a single element or family of elements or on certain types of reactions.

Physical chemistry. Physical chemistry is the branch of chemistry that investigates the physical properties of materials and relates these properties to the structure of the substance. Physical chemists study both

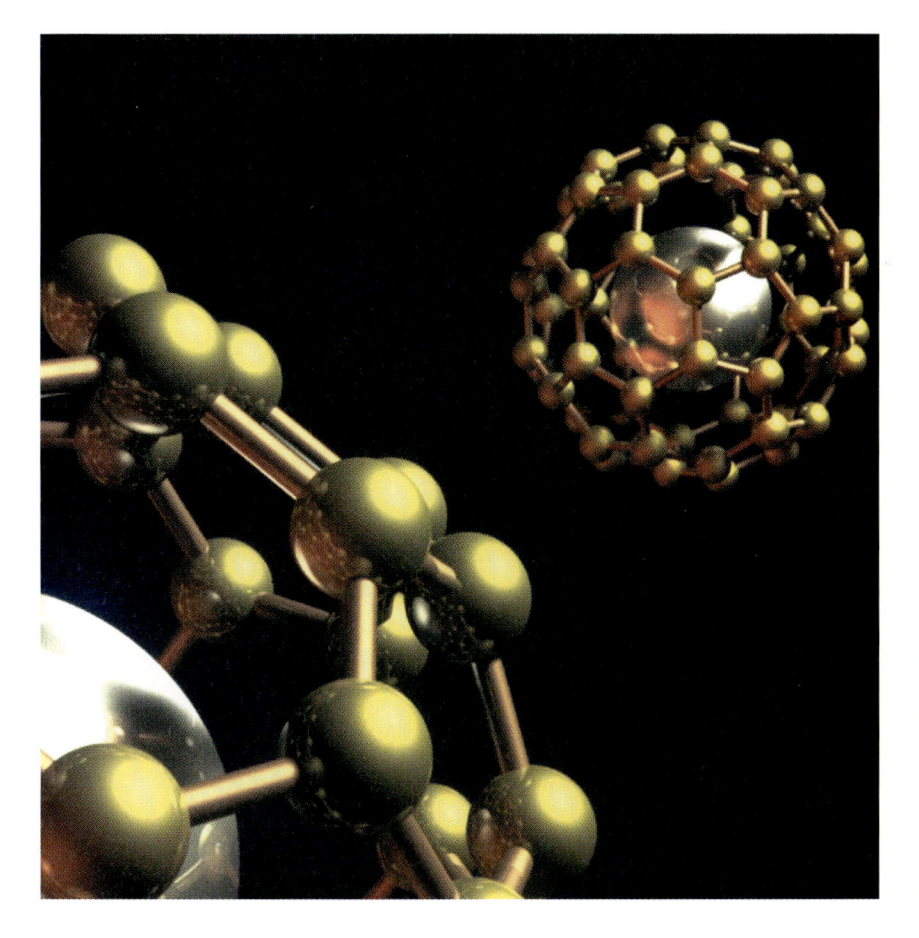

Computer-generated model of a 60-carbon molecule enclosing a potassium ion. The 60-carbon molecule, called a buckminsterfullerene, was discovered by organic chemists in 1985. *(Reproduced by permission of Photo Researchers, Inc.)*

Chemistry

organic and inorganic compounds and measure such variables as the temperature needed to liquefy a solid, the energy of the light absorbed by a substance, and the heat required to accomplish a chemical transformation. A computer is used to calculate the properties of a material and compare these assumptions to laboratory measurements. Physical chemistry is responsible for the theories and understanding of the physical phenomena utilized in organic and inorganic chemistry.

Analytical chemistry. Analytical chemistry is that field of chemistry concerned with the identification of materials and with the determination of the percentage composition of compounds and mixtures. These two lines of research are known, respectively, as qualitative analysis and quantitative analysis. Two of the oldest techniques used in analytical chemistry are gravimetric and volumetric analysis. Gravimetric analysis refers to the process by which a substance is precipitated (changed to a solid) out of solution and then dried and weighed. Volumetric analysis involves the reaction between two liquids in order to determine the composition of one or both of the liquids.

In the last half of the twentieth century, a number of mechanical systems have been developed for use in analytical research. For example, spectroscopy is the process by which an unknown sample is excited (or energized) by heating or by some other process. The radiation given off by the hot sample can then be analyzed to determine what elements are present. Various forms of spectroscopy are available (X-ray, infrared, and ultraviolet, for example) depending on the form of radiation analyzed.

Other analytical techniques now in use include optical and electron microscopy, nuclear magnetic resonance (MRI; used to produce a three-dimensional image), mass spectrometry (used to identify and find out the mass of particles contained in a mixture), and various forms of chromatography (used to identify the components of mixtures).

Other fields of chemistry. The division of chemistry into four major fields is in some ways misleading and inaccurate. In the first place, each of these four fields is so large that no chemist is an authority in any one field. An inorganic chemist might specialize in the chemistry of sulfur, the chemistry of nitrogen, the chemistry of the inert gases, or in even more specialized topics.

Secondly, many fields have developed within one of the four major areas, and many other fields cross two or more of the major areas. For an example of specialization, the subject of biochemistry is considered a subspecialty of organic chemistry. It is concerned with organic compounds that occur within living systems. An example of a cross-discipline subject is bioinorganic chemistry. Bioinorganic chemistry is the science dealing with the role of inorganic elements and their compounds (such as iron, copper, and sulfur) in living organisms.

At present, chemists explore the boundaries of chemistry and its connections with other sciences, such as biology, environmental science, geology, mathematics, and physics. A chemist today may even have a so-called nontraditional occupation. He or she may be a pharmaceutical salesperson, a technical writer, a science librarian, an investment broker, or a patent lawyer, since discoveries by a traditional chemist may expand and diversify into a variety of fields that encompass our whole society.

[*See also* **Alchemy; Mass spectrometry; Organic chemistry; Qualitative analysis; Quantitative analysis; Spectroscopy**]

Cholesterol

Cholesterol is a waxy substance found in the blood and body tissues of animals. It is an important structural component of animal cell membranes. Cholesterol is a lipid, a group of fats or fatlike compounds that do not dissolve in water. More specifically, it is a type of lipid known as a steroid. Other steroids include hormones, which are chemical substances produced by the body that regulate certain activities of cells or organs.

Cholesterol in the human body

Cholesterol is a biologically important compound in the human body. It is produced by the liver and used in the manufacture of vitamin D, adrenal gland hormones, and sex hormones. Large concentrations of cholesterol are found in the brain, spinal cord, and liver. Gallstones that occur in the gall bladder are largely made up of cholesterol. It is also found in bile (a fluid secreted by the liver), from which it gets its name: *chol* (Greek for "bile") plus *stereos* (Greek for "solid").

Normally, cholesterol produced by the liver circulates in the blood and is taken up by the body's cells for their needs. Cholesterol can also be removed from the blood by the liver and secreted in bile into the small intestine. From the intestine, cholesterol is released back into the bloodstream.

The body does not need cholesterol from dietary sources because the liver makes cholesterol from other nutrients. Eating saturated fats can cause the liver to produce more cholesterol than the body needs. Therefore, a diet high in saturated fats and cholesterol can raise blood

Cholesterol

> **Words to Know**
>
> **Atherosclerosis:** A disease in which plaques composed of cholesterol and fatty material form on the walls of arteries.
>
> **Bile:** A fluid secreted by the liver that aids in the digestion of fats and oils in the body.
>
> **High-density lipoprotein (HDL):** A lipoprotein low in cholesterol that is thought to protect against atherosclerosis.
>
> **Lipoprotein:** A large molecule composed of a lipid (a fat or fatlike compound), such as cholesterol, and a protein.
>
> **Low-density lipoprotein (LDL):** A lipoprotein high in cholesterol that is associated with increased risk of atherosclerosis.
>
> **Proteins:** Large molecules that are essential to the structure and functioning of all living cells.
>
> **Saturated fat:** Fats that are solid at room temperature or that become hard when exposed to cold temperatures.

cholesterol levels. Excess cholesterol that is not taken up by body cells may be deposited in the walls of arteries.

Cholesterol and heart disease. There has been much debate in the scientific community concerning the relationship between eating foods high in cholesterol and developing atherosclerosis (the blockage of coronary arteries with deposits of fatty material). Atherosclerosis impairs the flow of blood through arteries and leads to heart disease. A high blood cholesterol level is a risk factor for coronary artery disease.

Studies have shown that the major dietary cause of increased blood cholesterol levels is eating foods high in saturated fats (found mostly in animal products)—not foods containing cholesterol, as was once believed. Smoking, lack of exercise, obesity, caffeine, and heredity are other factors influencing blood cholesterol levels.

"Good" cholesterol and "bad" cholesterol

Cholesterol is carried in the blood bound to protein molecules called lipoproteins. Most of the cholesterol is transported on low-density lipoproteins (LDLs). LDL receptors on body cell membranes help regulate the blood cholesterol level by binding with LDLs, which are then taken up by the cells. However, if there are more LDLs than LDL receptors, the excess LDLs, or "bad" cholesterol, can be deposited in the lining of the arteries. High-density lipoproteins (HDLs), or "good" cholesterol, are thought to help protect against damage to the artery walls by carrying excess LDL back to the liver.

[*See also* **Circulatory system; Heart; Lipid; Nervous system**]

Cholesterol

A false color scanning electron micrograph of crystals of cholesterol. *(Reproduced by permission of Photo Researchers, Inc.)*

Chromosome

A chromosome is a structure that occurs within cells and that contains the cell's genetic material. That genetic material, which determines how an organism develops, is a molecule of deoxyribonucleic acid (DNA). A molecule of DNA is a very long, coiled structure that contains many identifiable subunits known as genes.

In prokaryotes, or cells without a nucleus, the chromosome is merely a circle of DNA. In eukaryotes, or cells with a distinct nucleus, chromosomes are much more complex in structure.

Historical background

The terms chromosome and gene were used long before biologists really understood what these structures were. When the Austrian monk and biologist Gregor Mendel (1822–1884) developed the basic ideas of heredity, he assumed that genetic traits were somehow transmitted from parents to offspring in some kind of tiny "package." That package was later given the name "gene." When the term was first suggested, no one had any idea as to what a gene might look like. The term was used simply to convey the idea that traits are transmitted from one generation to the next in certain discrete units.

Magnification of chromosome 17, which carries the breast and ovarian cancer gene. *(Reproduced by permission of Custom Medical Stock Photo, Inc.)*

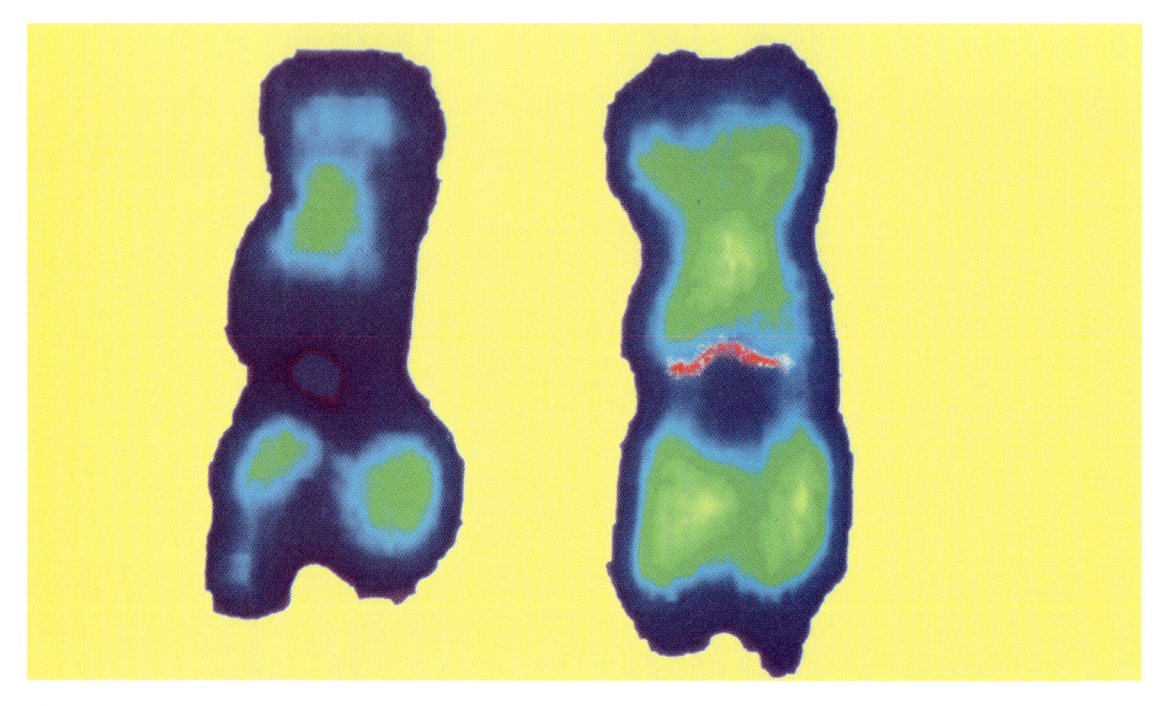

Chromosome

> **Words to Know**
>
> **Deoxyribonucleic acid (DNA):** The genetic material in the nucleus of cells that contains information for an organism's development.
>
> **Eukaryote:** A cell with a distinct nucleus.
>
> **Nucleotide:** The building blocks of nucleic acids.
>
> **Prokaryote:** A cell without a nucleus.
>
> **Protein:** Large molecules that are essential to the structure and functioning of all living cells.

The term "chromosome" was first suggested in 1888 by the German anatomist Heinrich Wilhelm Gottfried von Waldeyer-Hartz (1836–1921). Waldeyer-Hartz used the term to describe certain structures that form during the process of cell division (reproduction).

One of the greatest breakthroughs in the history of biology occurred in 1953 when American biologist James Watson (1928–) and English chemist Francis Crick (1916–) discovered the chemical structure of a class of compounds known as deoxyribonucleic acids (DNA). The Watson and Crick discovery made it possible to express biological concepts (such as the gene) and structures (such as the chromosome) in concrete chemical terms.

The structure of chromosomes and genes

Today we know that a chromosome contains a single molecule of DNA along with several kinds of proteins. A molecule of DNA, in turn, consists of thousands and thousands of subunits, known as nucleotides, joined to each other in very long chains. A single molecule of DNA within a chromosome may be as long as 8.5 centimeters (3.3 inches). To fit within a chromosome, the DNA molecule has to be twisted and folded into a very complex shape.

Imagine that a DNA molecule is represented by a formula such as this:

$$-[-N_1-N_4-N_2-N_2-N_2-N_1-N_3-N_2-N_3-N_4-N_1-N_2-N_3-N_3-N_1-N_1-N_2-N_3-N_4-]-$$

In this formula, the abbreviations N_1, N_2, N_3, and N_4 stand for the four different nucleotides used in making DNA. The brackets at the beginning

Chromosome

and end of the formula mean that the actual formula goes on and on. A typical molecule of DNA contains up to three billion nucleotides. The unit shown above, therefore, is no more than a small portion of the whole DNA molecule.

Each molecule of DNA can be subdivided into smaller segments consisting of a few thousand or a few tens of thousands of nucleotides. Each of these subunits is a gene. Another way to represent a DNA molecule, then, is as follows:

-[-G-D-N-E-Y-D-A-B-W-Q-X-C-R-K-S-]-

where each different letter stands for a different gene.

The function of genes and chromosomes

Each gene in a DNA molecule carries the instructions for making a single kind of protein. Proteins are very important molecules that perform many vital functions in living organisms. For example, they serve as hormones, carrying messages from one part of the body to another part; they act as enzymes, making possible chemical reactions that keep the cell alive; and they function as structural materials from which cells can be made.

Every cell has certain specific functions to perform. The purpose of a bone cell, for example, is to make more bone. The purpose of a pancreas cell, on the other hand, might be to make the compound insulin, which aids in the manufacture of glucose (blood sugar).

The job of genes in a DNA molecule, therefore, is to tell cells how to manufacture all the different chemical compounds (proteins) they need to make in order to function properly. The way in which they carry out this function is fairly straightforward. At one point in the cell's life, its chromosomes become untangled and open up to expose their genes. The genes act as a pattern from which proteins can be built. The proteins that are constructed in the cell are determined, as pointed out above, by the instructions built into the gene.

When the proteins are constructed, they are released into the cell itself or into the environment outside the cell. They are then able to carry out the functions for which they were made.

Chromosome numbers and Xs and Ys

Each species has a different number of chromosomes in their nuclei. The mosquito, for instance, has 6 chromosomes. Lilies have 24, earthworms 36, chimps 48, and horses 64. The largest number of chromosomes

Chromosome

are found in the Adder's tongue fern, which has more than 1,000 chromosomes. Most species have, on average, 10 to 50 chromosomes. With 46 chromosomes, humans fall well within this average.

The 46 human chromosomes are arranged in 23 pairs. One pair of the 23 constitute the sex hormones, called the X and Y chromosomes. Males have both an X and a Y chromosome, while females have two X chromosomes. If a father passes on a Y chromosome, then his child will be male. If he passes on an X chromosome, then the child will be female.

A scanning electron micrograph of a human X chromosome. *(Reproduced by permission of Photo Researchers, Inc.)*

Cigarette smoke

The X chromosome is three times the size of the Y chromosome and carries 100 times the genetic information.

However, in 2000, scientists announced that the X and Y chromosomes were once a pair of identical twins. These identical chromosomes were found some 300 million years ago in reptiles, long before mammals arose. The genes in these creatures did not decide sex on their own. They responded to some environmental cue like temperature. That still goes on today in the eggs of turtles and crocodiles. But in a single animal at that time long ago, a mutation occurred on one of the pair of identical chromosomes, creating what scientists recognize today as the Y chromosome—a gene that when present always produces a male.

[*See also* **Genetic disorders; Genetic engineering; Genetics; Mendelian laws of inheritance; Molecular biology; Mutation; Nucleic acid; Protein**]

Cigarette smoke

Cigarette smoke contains cancer-causing substances called carcinogens. Cigarette smoking is the major cause of lung cancer and emphysema (a serious disease of the lungs). People who smoke are also at increased risk for developing other cancers, heart disease, and chronic lung ailments. In the United States alone, cigarette smoking is responsible for almost 500,000 premature deaths per year.

Cigarette smoke is called mainstream smoke when it is inhaled directly from a cigarette. Sidestream smoke is smoke that is emitted from a burning cigarette and exhaled from a smoker's lungs. Sidestream smoke is also called environmental tobacco smoke or secondhand smoke. Passive smoking, or the inhaling of secondhand smoke by nonsmokers, is believed to be responsible for about 3,000 lung cancer deaths per year. Nonsmokers exposed to secondhand smoke also have a greater chance of suffering from respiratory disorders.

Components of cigarette smoke

Over 4,000 different chemicals are present in cigarette smoke. Many of these are carcinogenic, or capable of causing changes in the genetic material of cells that can lead to cancer. Cigarette smoke contains nicotine, an addictive chemical, and carcinogenic tars. In addition, smoking produces carbon monoxide, which has the effect of decreasing the amount of oxygen in the blood.

When cigarette smoke is inhaled, the chemicals contained in it are quickly absorbed by the lungs and released into the bloodstream. From the blood, these chemicals pass into the brain, heart, kidneys, liver, lungs, gastrointestinal tract, muscle, and fat tissue. In pregnant women, cigarette smoke crosses the placenta and may affect development of the fetus.

The health consequences of smoking

There is a strong relationship between the length of time a person smokes, the number of cigarettes a person smokes each day, and the development of smoking-related diseases. Simply put, the more one smokes, the more one is likely to suffer ill effects.

Cigarette smoke weakens blood vessel walls and increases the level of cholesterol in the blood, which can lead to atherosclerosis (a disease in which fatty material is deposited in the arterial walls). It can cause the coronary arteries to narrow, increasing the risk of heart attack due to impaired blood flow to the heart. Smoking also increases the risk of stroke (a blood clot or rupture in an artery of the brain).

In addition to lung cancer, smoking can cause cancers of the mouth, throat, voicebox, esophagus, stomach, cervix, and bladder. Drinking alcohol while smoking causes 75 percent of all mouth and throat cancers.

Cigarette smoke

A normal lung (left) and the lung of a cigarette smoker. *(Reproduced by permission of Photo Researchers, Inc.)*

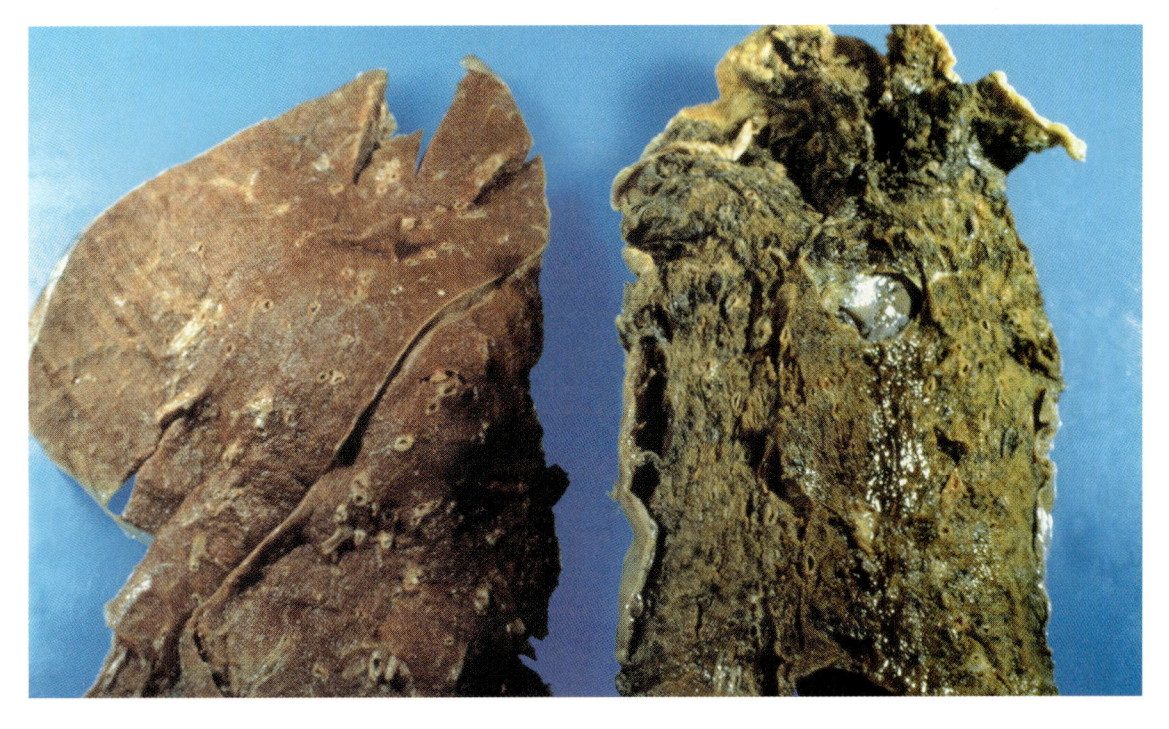

> ### Words to Know
>
> **Addiction:** Compulsive use of a habit-forming substance.
>
> **Carcinogen:** Any substance that is capable of causing cancer.
>
> **Dopamine:** A chemical in the brain that is associated with feelings of pleasure.
>
> **Nicotine:** A poisonous chemical that is the addictive substance in cigarettes.
>
> **Secondhand smoke:** The smoke emitted from a burning cigarette and exhaled from a smoker's lungs.

People who have a tendency to develop cancer because of hereditary factors may develop the disease more quickly if they smoke.

Smoking is the leading cause of lung disease in the United States and results in deaths from pneumonia, influenza, bronchitis, emphysema, and chronic airway obstruction. Smoking increases mucus production in the lungs and destroys cilia, the tiny hairlike structures that normally sweep debris out of the lungs.

Nicotine addiction

The nicotine in cigarette smoke causes the release of a chemical in the brain called dopamine. When the level of dopamine in the brain is increased, a person experiences feelings of extreme pleasure and contentment. In order to sustain these feelings, the level of nicotine in the body must remain constant; a smoker becomes dependent on the good feelings caused by the release of dopamine and thus becomes addicted to nicotine.

[*See also* **Addiction; Respiratory system**]

Circle

A circle can be defined as a closed curved line on which every point is equally distant from a fixed point within it. Following is some of the terminology used in referring to a circle:

1. The fixed point is called the *center* of the circle (C in Figure 1).

2. A line segment joining the center to any point on the circle is the *radius* of the circle (CA in Figure 1).

3. A line segment passing through the center of the circle and joining any two points on the circle is the *diameter* of the circle (DB in Figure 1). The diameter of a circle is twice its radius.

4. The distance around the circle is called the *circumference* of the circle.

5. Any portion of the curved line that makes up the circle is an *arc* of the circle (for example, AB or DA in Figure 1).

6. A straight line inside the circle joining the two end points of an arc is a *chord* of the circle (DE in Figure 1).

Mathematical relationships

One of the interesting facts about circles is that the ratio between their circumference and their diameter is always the same, no matter what size the circle is. That ratio is given the name pi (π) and has the value of 3.141592+. Pi is an irrational number. That is, it cannot be expressed as the ratio of two whole numbers. The + added at the end of the value above means that the value of pi is indeterminate: you can continue to divide the circumference of any circle by its diameter forever and never get an answer without a remainder.

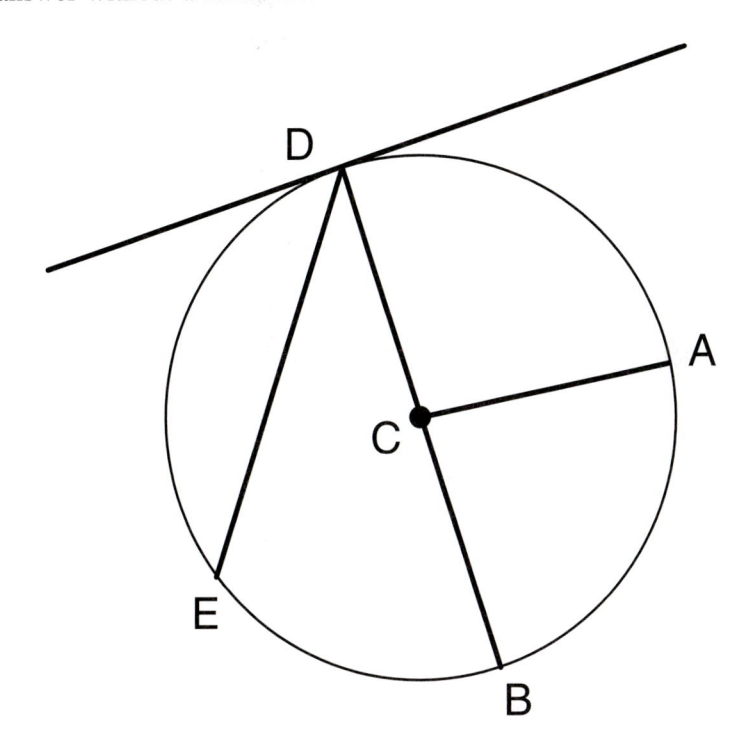

Figure 1. A circle. *(Reproduced by permission of The Gale Group.)*

Circulatory system

The area of any circle is equal to its radius squared multiplied by π, or: $A = \pi r^2$. The circumference of a circle can be found by multiplying its diameter by π ($C = \pi D$) or twice its radius by π ($C = 2\pi r$).

Circulatory system

The human circulatory system is responsible for delivering food, oxygen, and other needed substances to all cells in all parts of the body while taking away waste products. The circulatory system is also known as the cardiovascular system, from the Greek word *kardia,* meaning "heart," and the Latin *vasculum,* meaning "small vessel." The basic components of the cardiovascular system are the heart, the blood vessels, and the blood. As blood circulates around the body, it picks up oxygen from the lungs, nutrients from the small intestine, and hormones from the endocrine glands, and delivers these to the cells. Blood then picks up carbon dioxide and cellular wastes from cells and delivers these to the lungs and kidneys, where they are excreted.

The human heart

The adult heart is a hollow cone-shaped muscular organ located in the center of the chest cavity. The lower tip of the heart tilts toward the left. The heart is about the size of a clenched fist and weighs approximately 10.5 ounces (300 grams). A heart beats more than 100,000 times a day and close to 2.5 billion times in an average lifetime. The pericardium—a triple-layered sac—surrounds, protects, and anchors the heart. Pericardial fluid located in the space between two of the layers reduces friction when the heart moves.

The heart is divided into four chambers. A septum or partition divides it into a left and right side. Each side is further divided into an upper and lower chamber. The upper chambers, the atria (singular atrium), are thin-walled. They receive blood entering the heart and pump it to the ventricles, the lower heart chambers. The walls of the ventricles are thicker and contain more cardiac muscle than the walls of the atria. This enables the ventricles to pump blood out to the lungs and the rest of the body.

The left and right sides of the heart function as two separate pumps. The right atrium receives blood carrying carbon dioxide from the body through a major vein, the vena cava, and delivers it to the right ventricle. The right ventricle, in turn, pumps the blood to the lungs via the pulmonary artery. The left atrium receives the oxygen-rich blood from the

> ### Words to Know
>
> **Artery:** Vessel that transports blood away from the heart.
>
> **Atherosclerosis:** Condition in which fatty material such as cholesterol accumulates on artery walls forming plaque that obstructs blood flow.
>
> **Atrium:** Receiving chamber of the heart.
>
> **Capillary:** Vessel that connects artery to vein.
>
> **Diastole:** Period of relaxation and expansion of the heart when its chambers fill with blood.
>
> **Hormones:** Chemical messengers that regulate body functions.
>
> **Hypertension:** High blood pressure.
>
> **Sphygmomanometer:** Instrument that measures blood pressure in millimeters of mercury.
>
> **Systole:** Rhythmic contraction of the heat.
>
> **Vein:** Vessel that transports blood to the heart.
>
> **Ventricle:** Pumping chamber of the heart.

lungs from the pulmonary veins, and delivers it to the left ventricle. The left ventricle then pumps it into the aorta, the major artery that leads to all parts of the body. The wall of the left ventricle is thicker than the wall of the right ventricle, making it a more powerful pump, able to push blood through its longer trip around the body.

One-way valves in the heart keep blood flowing in the right direction and prevent backflow. The valves open and close in response to pressure changes in the heart. Atrioventricular valves are located between the atria and ventricles. Semilunar valves lie between the ventricles and the major arteries into which they pump blood. People with a heart murmur have a defective heart valve that allows the backflow of blood.

The heart cycle refers to the events that occur during a single heartbeat. The cycle involves systole (the contraction phase) and diastole (the relaxation phase). In the heart, the two atria contract while the two ventricles relax. Then, the two ventricles contract while the two atria relax. The heart cycle consists of a systole and diastole of both the atria and ventricles. At the end of a heartbeat all four chambers rest. The average

Circulatory system

heart beats about 75 times per minute, and each heart cycle takes about 0.8 seconds.

Blood vessels

The blood vessels of the body (arteries, capillaries, and veins) make up a closed system of tubes that carry blood from the heart to tissues all over the body and then back to the heart. Arteries carry blood away from the heart, while veins carry blood toward the heart. Large arteries leave

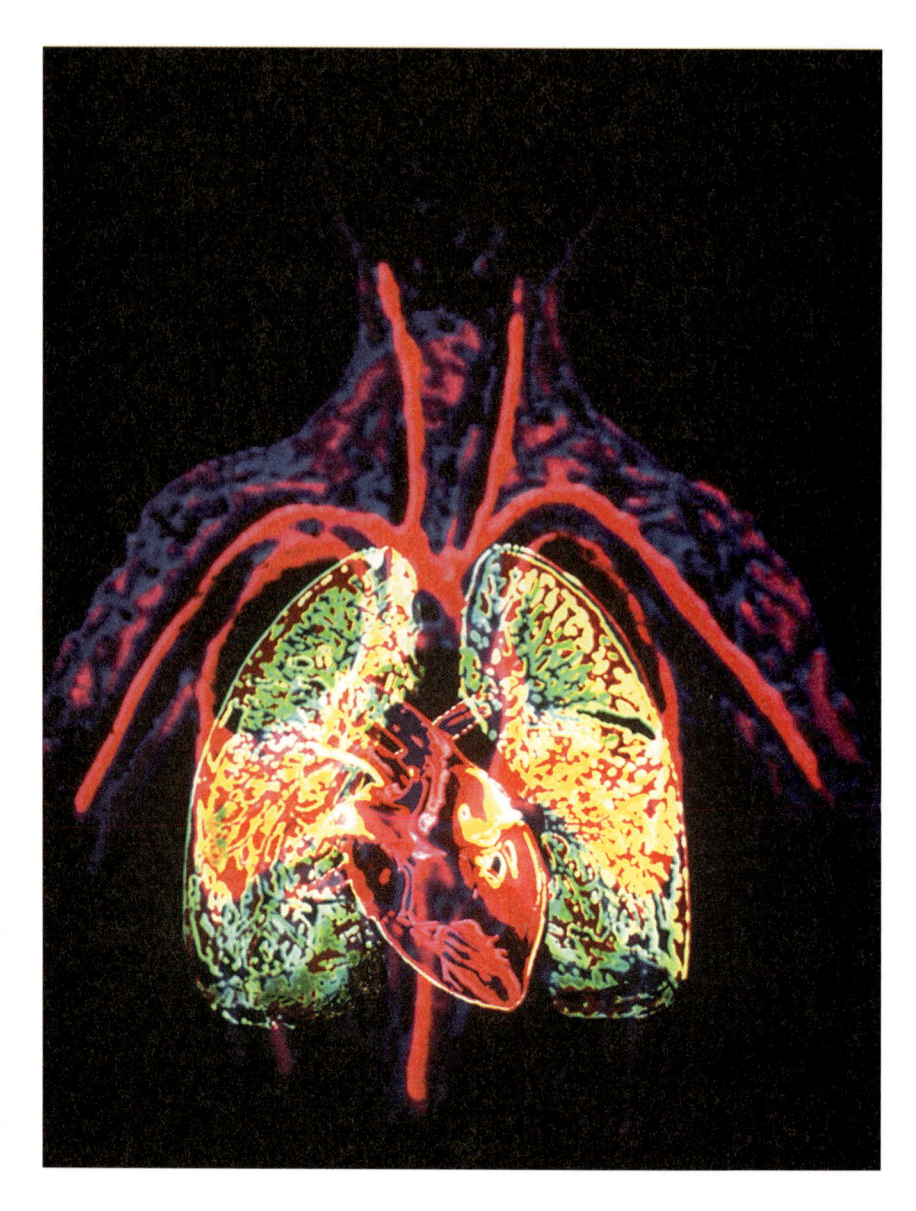

An image of the main components of the human circulatory system. The heart (placed between the lungs) delivers blood to the lungs, where it picks up oxygen and circulates it throughout the body by means of a system of blood vessels. *(Reproduced by permission of The Stock Market.)*

482   U·X·L Encyclopedia of Science, 2nd Edition

the heart and branch into smaller ones that reach out to various parts of the body. These divide still further into smaller vessels called arterioles that penetrate the body tissues. Within the tissues, the arterioles branch into a network of microscopic capillaries. Substances move in and out of the capillary walls as the blood exchanges materials with the cells. Before leaving the tissues, capillaries unite into venules, which are small veins. The venules merge to form larger and larger veins that eventually return blood to the heart.

The walls of arteries, veins, and capillaries differ in structure. In all three, the vessel wall surrounds a hollow center through which the blood flows. The walls of both arteries and veins are composed of three coats, but they differ in thickness. The inner and middle coats of arteries are thicker than those of veins. This makes arteries more elastic and capable of expanding when blood surges through them from the beating heart. The walls of veins are more flexible than artery walls. This allows skeletal muscles to contract against them, squeezing the blood along as it returns to the heart. One-way valves in the walls of veins keep blood flowing in one direction. The walls of capillaries are only one cell thick. Of all the blood vessels, only capillaries have walls thin enough to allow the exchange of materials between cells and the blood.

Blood pressure is the pressure of blood against the wall of an artery. Blood pressure originates when the ventricles contract during the heartbeat. It is strongest in the aorta and decreases as blood moves through progressively smaller arteries. A sphygmomanometer (pronounced sfig-moe-ma-NOM-i-ter) is an instrument that measures blood pressure in millimeters (mm) of mercury. Average young adults have a normal blood pressure reading of about 120 mm for systolic pressure and 80 mm for diastolic pressure. Blood pressure normally increases with age.

Blood

Blood is liquid connective tissue. It transports oxygen from the lungs and delivers it to cells. It picks up carbon dioxide from the cells and brings it to the lungs. It carries nutrients from the digestive system and hormones from the endocrine glands to the cells. It takes heat and waste products away from cells. It protects the body by clotting and by fighting disease through the immune system.

Blood is heavier and stickier than water, and has a temperature in the body of about 100.4°F (38°C). Blood makes up approximately 8 percent of an individual's total body weight. A male of average weight has about 1.5 gallons (5.5 liters) of blood in his body, while a female has about 1.2 gallons (4.5 liters).

Clone and cloning

Blood is composed of plasma (liquid portion) and blood cells. Plasma, which is about 91.5 percent water, carries blood cells and helps conduct heat. The three types of cells in blood are red blood cells (erythrocytes), white blood cells (leukocytes), and platelets (thrombocytes). More than 99 percent of all the blood cells are red blood cells. They contain hemoglobin, a red pigment that carries oxygen, and each red cell has about 280 million hemoglobin molecules. White blood cells fight disease organisms by destroying them or by producing antibodies. Platelets bring about clotting of the blood.

Circulatory diseases. Two disorders that involve blood vessels are hypertension and atherosclerosis. Hypertension, or high blood pressure, is the most common circulatory disease. In about 90 percent of hypertension sufferers, blood pressure stays high without any known physical cause. Limiting salt and alcohol intake, stopping smoking, losing weight, increasing exercise, and managing stress all help reduce blood pressure. Medications also help control hypertension.

In atherosclerosis, fatty material such as cholesterol accumulates on the artery wall forming plaque that obstructs blood flow. The plaque can form a clot that breaks off, travels in the blood, and can block a smaller vessel. A stroke may occur when a clot obstructs an artery or capillary in the brain. Treatment for atherosclerosis includes medication, surgery, a high-fiber diet low in fat, and exercise.

[*See also* **Blood; Heart; Lymphatic system**]

Clone and cloning

A clone is a cell, group of cells, or organism produced by asexual reproduction that contains genetic information identical to that of the parent cell or organism. Asexual reproduction is the process by which a single parent cell divides to produce two new daughter cells. The daughter cells produced in this way have exactly the same genetic material as that contained in the parent cell.

Although some organisms reproduce asexually naturally, the term "cloning" today usually refers to artificial techniques for achieving this result. The first cloning experiments conducted by humans involved the growth of plants that developed from grafts and stem cuttings. Modern cloning practices that involve complex laboratory techniques is a relatively recent scientific advance that is at the forefront of modern biology.

Clone and cloning

> **Words to Know**
>
> **DNA (deoxyribonucleic acid):** The specific molecules that contain genetic information in an organism.
>
> **Embryo:** The earliest stage of animal development in the uterus before the animal is considered a fetus.
>
> **Genes:** Specific biological components that carry the instructions for the formation of an organisms and its specific traits, such as eye or hair color.
>
> **Genetic engineering:** The process of combining specific genes to attain desired traits.
>
> **Genetics:** The study of hereditary traits passed on through the genes.
>
> **Heredity:** Characteristics passed on from parents to offspring.
>
> **Nucleus:** Plural is nuclei; the part of the cell that contains most of its genetic material, including chromosomes and DNA.

Among these techniques is the ability to isolate and make copies of (clone) individual genes that direct an organism's development. Cloning has many promising applications in medicine, industry, and basic research.

History of cloning

Humans have used simple methods of cloning such as grafting and stem cutting for more than 2,000 years. The modern era of laboratory cloning began in 1958 when the English-American plant physiologist Frederick C. Steward (1904–1993) cloned carrot plants from mature single cells placed in a nutrient culture containing hormones, chemicals that play various and significant roles in the body.

The first cloning of animal cells took place in 1964. In the first step of the experiment, biologist John B. Gurdon first destroyed with ultraviolet light the genetic information stored in a group of unfertilized toad eggs. He then removed the nuclei (the part of an animal cell that contains the genes) from intestinal cells of toad tadpoles and injected them into those eggs. When the eggs were incubated (placed in an environment that promotes growth and development), Gurdon found that 1 to 2 percent of the eggs developed into fertile, adult toads.

Clone and cloning

The first successful cloning of mammals was achieved nearly 20 years later. Scientists in both Switzerland and the United States successfully cloned mice using a method similar to that of Gurdon. However, the Swiss and American methods required one extra step. After the nuclei were taken from the embryos of one type of mouse, they were transferred into the embryos of another type of mouse. The second type of mouse served as a surrogate (substitute) mother that went through the birthing process to create the cloned mice. The cloning of cattle livestock was achieved in 1988 when embryos from prize cows were transplanted to unfertilized cow eggs whose own nuclei had been removed.

Dolly. All of the above experiments had one characteristic in common: they involved the use of embryonic cells, cells at a very early stage of development. Biologists have always believed that such cells have the ability to adapt to new environments and are able to grow and develop in a

Dolly, Ian Wilmut's sheep clone, at eight months old. (Reproduced by permission of Photo Researchers, Inc.)

cell other than the one from which they are taken. Adult cells, they have thought, do not retain the same adaptability.

A startling announcement in February 1997 showed the error in this line of reasoning. The Scottish embryologist Ian Wilmut (1945–) reported that he had cloned an adult mammal for the first time. The product of the experiment was a sheep named Dolly, seven months old at the time of the announcement.

In Wilmut's experiment, the nucleus from a normal embryonic cell from an adult sheep was removed. A cell from another adult sheep's mammary gland was then removed and transferred to the empty cell from the first sheep. The embryonic cell began to grow normally and a young sheep (Dolly) was eventually born. A study of Dolly's genetic make-up has shown that she is identical to the second sheep, the adult female that supplied the genetic material for the experiment.

Rapid advances in cloning

Advances in the cloning process have developed rapidly since Dolly made her debut. Only a year and a half after Dolly was cloned, Ryuzo Yanagimachi, a biologist from the University of Hawaii, announced in July 1998 that he and his research team had made dozens of mouse clones and even cloned some of those that had been first cloned. What made the cloning of adult mice astounding is that mouse embryos develop quickly after fertilization. Scientist had thought a mouse would prove to be difficult or impossible to clone due to its embryonic development. That cloning of dozens of adult mice took place only a little more than a year after Dolly astounded the scientific world.

Later in 1998, a team of scientists led by Yukio Tsunoda from Kinki University in Japan announced that they had cloned eight calves from a single cow. Eighty percent of the embryos cloned survived until birth— an excellent efficiency rate. Later, four of the eight calves died from causes unrelated to cloning.

In January 2001, scientists in the United States announced they had cloned an endangered species, a baby Asian ox called a gaur (pronounced GOW-er). It was the scientific world's first attempt at replicating an endangered species. Scientists say such cloning could save endangered animals from extinction or even bring back species already extinct. To clone the gaur, the scientists removed the nucleus from a cow's egg cell and replaced it with the nucleus of a gaur skin cell. They then placed the fertilized egg cell in the womb of a domestic cow, which brought the gaur

Clone and cloning

to term. Unfortunately, just 48 hours after the gaur baby was born, it died of dysentery (diarrhea), which the scientists believed was not related to the cloning. Undeterred, the scientists stated they had long-term goals for more endangered species cloning research.

The cloning process

Simple organisms are relatively easy to clone. In some cases, entire cells can be inserted into bacteria or a yeast culture that reproduces asexually. As these cultures multiply, so do the cells inserted into them.

The cloning of higher animals is generally more difficult. One approach is to remove the nucleus of one cell by means of very delicate instruments and then to insert that nucleus into a second cell. Another method is to divide embryo tissues and insert them into surrogate mothers, where they then develop normally.

The benefits of cloning

The cloning of cells promises to produce many benefits in farming, medicine, and basic research. In the realm of farming, the goal is to clone plants that contain specific traits that make them superior to naturally occurring plants. For example, field tests have been conducted using clones of plants whose genes have been altered in the laboratory (by genetic engineering) to produce resistance to insects, viruses, and bacteria. New strains of plants resulting from the cloning of specific traits could also lead to fruits and vegetables with improved nutritional qualities and longer shelf lives, or new strains of plants that can grow in poor soil or even under water.

A cloning technique known as twinning could induce livestock to give birth to twins or even triplets, thus reducing the amount of feed needed to produce meat. And as was shown, cloning also holds promise for saving certain rare breeds of animals from extinction.

In the realm of medicine and health, gene cloning has been used to produce vaccines and hormones. Cloning techniques have already led to the inexpensive production of the hormone insulin for treating diabetes and of growth hormones for children who do not produce enough hormones for normal growth. The use of monoclonal antibodies in disease treatment and research involves combining two different kinds of cells (such as mouse and human cancer cells) to produce large quantities of specific antibodies. These antibodies are produced by the immune system to fight off disease. When injected into the blood stream, the cloned antibodies seek out and attack disease-causing cells anywhere in the body.

The ethics of cloning

Clone and cloning

The scientific world continues to be amazed by the speed of the development of cloning. Some scientists now suggest that the cloning of humans could occur in the near future. Despite the benefits of cloning and its many promising avenues of research, however, certain ethical questions concerning the possible abuse of cloning have been raised. At the heart of these questions is the idea of humans tampering with life in a way that could harm society, either morally or in a real physical sense. Some people object to cloning because it allows scientists to "act like God" in the manipulation of living organisms.

The cloning of Dolly raised the debate over this practice to a whole new level. It has become obvious that the technology for cloning Dolly could also be used to clone humans. A person could choose to make two or ten or a hundred copies of himself or herself by the same techniques used with Dolly. This realization has stirred an active debate about the morality of cloning humans. Some people see benefits from the practice, such as providing a way for parents to produce a new child to replace one dying of a terminal disease. Other people worry about humans taking into their own hands the future of the human race.

At the beginning of the twenty-first century, many scientists say the controversy over the ethics of cloning humans is exaggerated because of

Jars of identical banana plants in a cloning laboratory. These plants are ready to be subdivided and recultivated. *(Reproduced by permission of Photo Researchers, Inc.)*

the unpredictability of cloning in general. While scientists have cloned animals such as sheep, mice, cows, pigs, and goats (and have even made clones of clones on down for six generations), fewer than 3 percent of all those cloning efforts have succeeded. The animal clones that have been produced often have health problems—developmental delays, heart defects, lung problems, and malfunctioning immune systems. Scientists believe the breathtakingly rapid reprogramming in cloning can introduce random errors into a clone's DNA. Those errors have altered individual genes in minor ways, and the genetic defects have led to the development of major medical problems. Some scientists say this should make human cloning out of the question, but others counter that cloning humans may actually be easier and safer than cloning animals. Scientists agree that further research in the field of cloning is needed.

[*See also* **Genetic engineering; Nucleic acid; Reproduction**]

Clouds

Clouds are made up of minute water droplets or ice crystals that condense in the atmosphere. The creation of a cloud begins at ground level. As the Sun heats Earth's surface, the warmed ground heats the surrounding air, which then rises. This air contains variable amounts of water vapor that has evaporated from bodies of water and plants on Earth's surface. As the warmed ground-level air rises, it expands, cooling in the process. When the cooled air reaches a certain temperature, called the dew point, the water vapor in the air condenses into tiny microscopic droplets, forming a cloud. If condensation occurs below the freezing point (32°F; 0°C), ice crystals form the cloud. Clouds appear white because sunlight reflects off the water droplets. Thick clouds appear darker at the bottom because sunlight is partially blocked.

Classification

English scientist Luke Howard (1772–1864) developed a system to classify clouds in 1803. He grouped clouds into three major types: cumulus (piled up heaps and puffs), cirrus (fibrous and curly), and stratus (stretched out and layered). To further describe clouds, he combined these terms and added descriptive prefixes, such as alto (high) and nimbus (rain).

The International Cloud Classification presently recognizes ten forms of clouds, which are grouped into four height categories. Low-level clouds range from ground level to 6,500 feet (2,000 meters); mid-level from 6,500 to 20,000 feet (2,000 to 6,100 meters); high-level from 20,000 to 40,000 feet (6,100 to 12,200 meters); and vertical from 1,600 to 20,000 feet (490 to 6,100 meters).

Low-level clouds: Stratus, nimbostratus, stratocumulus. There are three forms of low-level clouds. Stratus clouds, the lowest, blanket the sky and usually appear gray. They form when a large moist air mass slowly rises and condenses. Fog is a stratus cloud at ground level. Nimbostratus clouds are thick, darker versions of stratus clouds. They usually produce continuous rain or snow. Stratocumulus clouds are large, grayish masses, spread out in a puffy layer. Sometimes they appear as rolls. If they are thick enough, stratocumulus will produce light precipitation.

Middle-level clouds: Altostratus, altocumulus. The two forms of mid-level clouds have the prefix "alto" added to their names. Altostratus clouds appear as a uniform blue or gray sheet covering all or almost all areas of the sky. The Sun or the Moon may be totally covered or shine through very weakly. These clouds are usually layered, with ice crystals at the top, ice and snow in the middle, and water droplets at the bottom. Altostratus clouds yield very light precipitation. Altocumulus are dense, fluffy white or grey balls or masses. When closely bunched together, they appear like fish scales across the sky: this effect is called a mackerel sky.

Cumulonimbus clouds. *(Reproduced by permission of Walter A. Lyons.)*

High-level clouds: Cirrus, cirrostratus, cirrocumulus. The three forms of high-level clouds are called cirrus or have the prefix "cirro" added to their names. Cirrus clouds, the highest, are made completely of ice crystals (or needles of ice) because they form where freezing temperatures prevail. Cirrus clouds are often called mares' tails because of their white, feathery or wispy appearance. Cirrostratus clouds are also made completely of ice crystals. They usually cover the sky as a thin veil or sheet of white. These clouds are responsible for the halos that occur around the Sun or the Moon. Cirrocumulus clouds, the least common clouds, are small roundish masses, often having a rippled appearance. These clouds usually cover a large area. They are made of either ice crystals or supercooled water droplets (droplets that stay in liquid form below the freezing point).

Vertical clouds: Cumulus, cumulonimbus. Two forms of clouds can extend thousands of feet in height. Flat-based cumulus clouds are vertically thick and appear puffy, like heaps of mashed potatoes or heads of cauliflower. They form when a column of warm air rises, expands, cools, and condenses. Low-level cumulus clouds generally indicate fair weather, but taller cumulus can produce moderate to heavy showers. Cumulonimbus clouds are thunderstorm clouds, rising in the air like a tower or mountain. The peak of a mature cumulonimbus resembles the flattened shape of an anvil. Because they often contain powerful updrafts and downdrafts, cumulonimbus can create violent storms of rain, hail, or snow.

[*See also* **Precipitation; Weather forecasting**]

Coal

Coal is a naturally occurring combustible material consisting primarily of the element carbon. It also contains low percentages of solid, liquid, and gaseous hydrocarbons and/or other materials, such as compounds of nitrogen and sulfur. Coal is usually classified into subgroups known as anthracite, bituminous, lignite, and peat. The physical, chemical, and other properties of coal vary considerably from sample to sample.

Origins of coal

Coal is often referred to as a fossil fuel. That name comes from the way in which coal was originally formed. When plants and animals die, they normally decay and are converted to carbon dioxide, water, and other

Coal

Words to Know

Anthracite: Hard coal; a form of coal with high heat content and a high concentration of pure carbon.

Bituminous: Soft coal; a form of coal with less heat content and pure carbon content than anthracite, but more than lignite.

British thermal unit (Btu): A unit for measuring heat content in the British measuring system.

Coke: A synthetic fuel formed by the heating of soft coal in the absence of air.

Combustion: The process of burning; a form of oxidation (reacting with oxygen) that occurs so rapidly that noticeable heat and light are produced.

Gasification: Any process by which solid coal is converted to a gaseous fuel.

Lignite: Brown coal; a form of coal with less heat content and pure carbon content than either anthracite or bituminous coal.

Liquefaction: Any process by which solid coal is converted to a liquid fuel.

Oxide: An inorganic compound whose only negative part is the element oxygen.

Peat: A primitive form of coal with less heat content and pure carbon content than any form of coal.

Strip mining: A method for removing coal from seams located near Earth's surface.

products that disappear into the environment. Other than a few bones, little remains of the dead organism.

At some periods in Earth's history, however, conditions existed that made other forms of decay possible. The bodies of dead plants and animals underwent only partial decay. The products remaining from this partial decay are coal, oil, and natural gas—the so-called fossil fuels.

To imagine how such changes may have occurred, consider the following possibility. A plant dies in a swampy area and is quickly covered with water, silt, sand, and other sediments. These materials prevent the plant debris from reacting with oxygen in the air and decomposing to carbon dioxide and water—a process that would occur under normal circumstances. Instead, anaerobic (pronounced an-nuh-ROBE-ik) bacteria (bacteria that do not require oxygen to live) attack the plant debris and convert it to simpler forms: primarily pure carbon and simple compounds of carbon and hydrogen (hydrocarbons).

The initial stage of the decay of a dead plant is a soft, woody material known as peat. In some parts of the world, peat is still collected from boggy areas and used as a fuel. It is not a good fuel, however, as it burns poorly and produces a great deal of smoke.

If peat is allowed to remain in the ground for long periods of time, it eventually becomes compacted. Layers of sediment, known as overburden, collect above it. The additional pressure and heat of the overburden gradually converts peat into another form of coal known as lignite or brown coal. Continued compaction by overburden then converts lignite into bituminous (or soft) coal and finally, into anthracite (or hard) coal.

Coal has been formed at many times in the past, but most abundantly during the Carboniferous Age (about 300 million years ago) and again during the Upper Cretaceous Age (about 100 million years ago).

Today, coal formed by these processes is often found layered between other layers of sedimentary rock. Sedimentary rock is formed when sand, silt, clay, and similar materials are packed together under heavy pressure. In some cases, the coal layers may lie at or very near Earth's surface. In other cases, they may be buried thousands of feet underground. Coal seams usually range from no more than 3 to 200 feet (1 to 60 meters) in thickness. The location and configuration of a coal seam determines the method by which the coal will be mined.

Composition of coal

Coal is classified according to its heating value and according to the percentage of carbon it contains. For example, anthracite contains the highest proportion of pure carbon (about 86 to 98 percent) and has the highest heat value (13,500 to 15,600 Btu/lb; British thermal units per pound) of all forms of coal. Bituminous coal generally has lower concentrations of pure carbon (from 46 to 86 percent) and lower heat values (8,300 to 15,600 Btu/lb). Bituminous coals are often subdivided on the basis of their heat value, being classified as low, medium, and high volatile bituminous and subbituminous. Lignite, the poorest of the true coals in terms of heat value (5,500 to 8,300 Btu/lb), generally contains about 46 to 60 percent pure carbon. All forms of coal also contain other elements present in living organisms, such as sulfur and nitrogen, that are very low in absolute numbers but that have important environmental consequences when coals are used as fuels.

Properties and reactions

By far the most important property of coal is that it burns. When the pure carbon and hydrocarbons found in coal burn completely, only two products are formed, carbon dioxide and water. During this chemical reaction, a relatively large amount of heat energy is released. For this reason, coal has long been used by humans as a source of energy for heating homes and other buildings, running ships and trains, and in many industrial processes.

Environmental problems associated with burning coal. The complete combustion of carbon and hydrocarbons described above rarely occurs in nature. If the temperature is not high enough or sufficient oxygen is not provided to the fuel, combustion of these materials is usually incomplete. During the incomplete combustion of carbon and hydrocarbons, other products besides carbon dioxide and water are formed. These products include carbon monoxide, hydrogen, and other forms of pure carbon, such as soot.

During the combustion of coal, minor constituents are also oxidized (meaning they burn). Sulfur is converted to sulfur dioxide and sulfur trioxide, and nitrogen compounds are converted to nitrogen oxides. The incomplete combustion of coal and the combustion of these minor constituents results in a number of environmental problems. For example, soot formed during incomplete combustion may settle out of the air and deposit an unattractive coating on homes, cars, buildings, and other structures. Carbon monoxide formed during incomplete combustion is a toxic gas and may cause illness or death in humans and other animals. Oxides of sulfur and nitrogen react with water vapor in the atmosphere and then settle out in the air as acid rain. (Acid rain is thought to be responsible for the destruction of certain forms of plant and animal—especially fish—life.)

In addition to these compounds, coal often contains a small percentage of mineral matter: quartz, calcite, or perhaps clay minerals. These components do not burn readily and so become part of the ash formed during combustion. This ash then either escapes into the atmosphere or is left in the combustion vessel and must be discarded. Sometimes coal ash also contains significant amounts of lead, barium, arsenic, or other elements. Whether airborne or in bulk, coal ash can therefore be a serious environmental hazard.

Coal

Coal mining

Coal is extracted from Earth using one of two major methods: subsurface or surface (strip) mining. Subsurface mining is used when seams of coal are located at significant depths below Earth's surface. The first step in subsurface mining is to dig vertical tunnels into the earth until the coal seam is reached. Horizontal tunnels are then constructed off the vertical tunnel. In many cases, the preferred way of mining coal by this method is called room-and-pillar mining. In room-and-pillar mining, vertical columns of coal (the pillars) are left in place as the coal around them is removed. The pillars hold up the ceiling of the seam, preventing it from collapsing on miners working around them. After the mine has been abandoned, however, those pillars may collapse, bringing down the ceiling of the seam and causing the collapse of land above the old mine.

Surface mining can be used when a coal seam is close enough to Earth's surface to allow the overburden to be removed easily and inexpensively. In such cases, the first step is to strip off all of the overburden in order to reach the coal itself. The coal is then scraped out by huge power shovels, some capable of removing up to 100 cubic meters at a time. Strip mining is a far safer form of coal mining for coal workers, but it presents a number of environmental problems. In most instances, an area that has been strip-mined is terribly scarred. Restoring the area to its

A coal seam in northwest Colorado. (Reproduced by permission of JLM Visuals.)

original state can be a long and expensive procedure. In addition, any water that comes in contact with the exposed coal or overburden may become polluted and require treatment.

Resources

Coal is regarded as a nonrenewable resource, meaning it is not replaced easily or readily. Once a nonrenewable resource has been used up, it is gone for a very long time into the future, if not forever. Coal fits that description, since it was formed many millions of years ago but is not being formed in significant amounts any longer. Therefore, the amount of coal that now exists below Earth's surface is, for all practical purposes, all the coal available for the foreseeable future. When this supply of coal is used up, humans will find it necessary to find some other substitute to meet their energy needs.

Large supplies of coal are known to exist (proven reserves) or thought to be available (estimated resources) in North America, Russia and other parts of the former Soviet Union, and parts of Asia, especially China and India. China produces the largest amount of coal each year, about 22 percent of the world's total, with the United States (19 percent), the former members of the Soviet Union (16 percent), Germany (10 percent), and Poland (5 percent) following.

China is also thought to have the world's largest estimated resources of coal, as much as 46 percent of all that exists. In the United States, the largest coal-producing states are Montana, North Dakota, Wyoming, Alaska, Illinois, and Colorado.

Uses

For many centuries, coal was burned in small stoves to produce heat in homes and factories. As the use of natural gas became widespread in the latter part of the twentieth century, coal oil and coal gas quickly became unpopular since they were somewhat smoky and foul smelling. Today, the most important use of coal, both directly and indirectly, is still as a fuel, but the largest single consumer of coal for this purpose is the electrical power industry.

The combustion of coal in power-generating plants is used to make steam, which, in turn, operates turbines and generators. For a period of more than 40 years beginning in 1940, the amount of coal used in the United States for this purpose doubled in every decade. Although coal is no longer widely used to heat homes and buildings, it is still used in industries such as paper production, cement and ceramic manufacture, iron and steel production, and chemical manufacture for heating and for steam generation.

Another use for coal is in the manufacture of coke. Coke is nearly pure carbon produced when soft coal is heated in the absence of air. In most cases, 1 ton of coal will produce 0.7 ton of coke in this process. Coke is valuable in industry because it has a heat value higher than any form of natural coal. It is widely used in steelmaking and in certain chemical processes.

Conversion of coal

A number of processes have been developed by which solid coal can be converted to a liquid or gaseous form for use as a fuel. Conversion has a number of advantages. In a liquid or gaseous form, the fuel may be easier to transport. Also, the conversion process removes a number of impurities from the original coal (such as sulfur) that have environmental disadvantages.

One of these conversion methods is known as gasification. In gasification, crushed coal is forced to react with steam and either air or pure oxygen. The coal is converted into a complex mixture of gaseous hydrocarbons with heat values ranging from 100 Btu to 1000 Btu. One day it may be possible to construct gasification systems within a coal mine, making it much easier to remove the coal (in a gaseous form) from its original seam.

In the process of liquefaction, solid coal is converted to a petroleum-like liquid that can be used as a fuel for motor vehicles and other applications. On the one hand, both liquefaction and gasification are attractive technologies in the United States because of its very large coal resources. On the other hand, the wide availability of raw coal means that expensive new technologies have been unable to compete economically with the natural product.

[*See also* **Carbon family; Petroleum; Pollution**]

Coast and beach

The coast and beach, where the continents meet the ocean, are ever-changing environments where the conflicting processes of erosion (wearing away) and sedimentation (building up) take place. Coast is the land that borders an ocean or large body of water. Beach refers to a much smaller land region, usually just the area directly affected by wave action.

> **Words to Know**
>
> **Emergent coast:** A coast that is formed when sea level declines and is characterized by wave-cut cliffs and formerly underwater beaches.
>
> **Longshore drift:** Movement of sand parallel to the shore, caused by slowing and breaking waves approaching the shore at an angle.
>
> **Submergent coast:** A coast that is formed when sea level rises and is characterized by drowned river valleys.

Coasts

Coasts are generally classified into two types: emergent and submergent. Emergent coasts are those that are formed when sea level declines. Areas that were once covered by the sea emerge and form part of the landscape. This new land area, which was once protected underwater, is now attacked by waves and eroded. If the new land is a cliff, waves may undercut it, eventually causing the top portions of the cliff to fall into the sea. When this happens, the beach is extended at its base. Along emergent coast shorelines the water level is quite shallow for some distance offshore. Much of the coast along California is emergent coast.

Submergent coasts are those that are formed when sea level rises, flooding formerly exposed land areas. Valleys near coastal areas that had been carved out by rivers become estuaries, or arms of the sea that extend inland to meet the mouth of a river, for example, Chesapeake Bay in Virginia and Maryland. Hilly terrains become collections of islands, such as those off the coast of Maine.

Beaches

Most of the sand and other sediments making up a beach are supplied by weathered and eroded rock from the mainland that is deposited by rivers at the coast. At the beach, wave action moves massive amounts of sand. As waves approach shallow water, they slow down because of friction with the bottom. They then become steeper and finally break. It is during this slowing and breaking that sand gets transported.

When a breaking wave washes up onto the beach, it does so at a slight angle, moving sand both toward and slightly down the beach. When

Coast and beach

the water sloshes back, it does so directly, without any angle. As a result, the water moves the sand along the beach in a zigzag pattern. This is called longshore drift. The magnitude and direction of longshore drift depends on the strength of the waves and the angle at which they approach, and these may vary with the season.

Barrier islands

A barrier island is a long, thin, sandy stretch of land that lies parallel to a mainland coast. Between the barrier island and the mainland is a calm, protected water body such as a lagoon or bay. If the coastline has a broad, gentle slope, strong waves and other ocean currents carry sand offshore and then deposit it, creating these islands. In the United States, most barrier islands are found along the Gulf Coast and the Atlantic Coast as far north as Long Island, New York.

Sand being moved by longshore drift and being replenished on beaches by eroding highlands is a natural, constant cycle. Beaches erode, however, when humans intervene in the cycle, often by building on coastal land. Two methods used to remedy beach erosion include pumping sand onto beaches from offshore and building breakwaters away from shore to stop longshore drift.

[*See also* **Erosion; Ocean; Tides**]

The coast of the Pacific Ocean in Boardman State Park, Oregon, an example of an emergent coast. *(Reproduced by permission of Photo Researchers, Inc.)*

Cocaine

Cocaine is a powerful drug that stimulates the body's central nervous system. Prepared from the leaves of the coca shrub that grows in South America, it increases the user's energy and alertness, reduces appetite and the need for sleep, and heightens feelings of pleasure. Although United States law makes its manufacture and use for nonmedical purposes illegal, many people are able to obtain it illegally.

A powerful stimulant

Aside from a few extremely limited medical uses, cocaine has no other purpose except to give a person an intense feeling of pleasure known as a "high." While this may not seem like such a bad thing, the great number of physical side effects that accompany that high, combined with the powerful psychological dependence it creates, makes it an extremely dangerous drug to take. As a very powerful stimulant, cocaine not only gives users more energy, it makes them feel confident and even euphoric (pronounced yew-FOR-ik)—meaning they are extremely elated or happy, usually for no reason. This feeling of elation and power makes users believe they can do anything, yet when this high wears off, they usually feel upset, depressed, tired, and even paranoid.

Cocaine has a very interesting history: It has gone from being considered a mild stimulant and then a wonder drug, to a harmless "recreational" drug, and finally to a powerfully addictive and very dangerous illegal drug. Although cocaine has, in fact, been all of these things at one time or another, we know it today to be an addictive drug that can wreck a person physically, mentally, and socially. It can also easily kill people.

History and European discovery

Cocaine is extracted from the leaves of the coca shrub (*Erythroxylum coca*), which grows in the tropical forests on the slopes of the Andes Mountains of Peru. A second species, *Erythroxylum novagranatense*, grows naturally in the drier mountainous regions of Columbia. For thousands of years, the native populations of those areas chewed the leaves of these plants to help them cope with the difficulty of living at such a high altitude. Chewing raw coca leaves (usually combined with ashes or lime) reduced their fatigue and suppressed their hunger, making them better able to handle the hard work they had to do to live so high up in the mountains. The coca leaves were also used during religious ceremonies and for rituals such as burials. The feelings that the leaves gave to their chewers made them consider the coca plant to be a gift from the gods.

Cocaine

> ### Words to Know
>
> **Coca leaves:** Leaves of the coca plant from which cocaine is extracted.
>
> **Crack:** A smokable and inexpensive form of pure cocaine sold in the form of small pellets or "rocks."
>
> **Euphoria:** A feeling of elation.

Once European explorers started coming to the Americas in the late fifteenth century, it was only a matter of time until invaders, such as the Spanish, came to the New World seeking riches. By the time the Spanish arrived in what is now Peru, the people of that land, known as the Incas, were already a civilization in decline, and they were easily subdued and conquered. The Spaniards eventually learned that giving coca leaves to native workers enabled them to force the workers to do enormous amounts of work in the gold and silver mines that were located in high altitudes. For the next two hundred years, although some coca plants were taken back to Europe, they were not popular or well-known since they did not travel well and were useless if dried out. Further, the Europeans did not like all the chewing and spitting required to get at the plant's active ingredient, and until this part of the plant could be isolated, coca leaves were not very much in demand.

Active part isolated

All of this changed by the middle of the nineteenth century when German physician Albert Niemann perfected the process of isolating the active part of the drug and improved the process of making it. Niemann extracted a purified form of cocaine from the coca leaves, and wrote about the anesthetic or numbing feeling obtained when he put it on his tongue. Cocaine then began its inevitable introduction into medicine, drink, and finally drug abuse. First it was considered by many doctors to be a wonder drug, and they began prescribing it for all sorts of physical and mental problems. By the 1880s, cocaine was even added to a very popular "medicinal" wine called Vin Mariani. The famous Austrian physician Sigmund Freud (1856–1939), who would become the founder of psychoanalysis, published a paper in 1884 that made many wrong medical claims for cocaine. Although he would later withdraw his claims, Freud did write at the time, "The use of coca in moderation is more likely to promote health than to impair it."

Popular use

In 1888, a soft drink named "Coca-Cola" was developed in America that contained cocaine and advertised itself as "the drink that relieves exhaustion." By 1908, however, the makers of Coca-Cola realized their mistake and removed all the cocaine from it, using only caffeine as a stimulant. By then, the initial enthusiasm for cocaine was seen to be undeserved, and many cases of overuse and dependence eventually forced lawmakers to take action against it. Consequently, in 1914 the United States introduced the Harrison Narcotic Act, which made cocaine illegal. After that, cocaine use was popular only with a fairly small number of artists, musicians, and the very rich, until the 1970s. In that decade, cocaine use skyrocketed as many young people who had earlier smoked marijuana

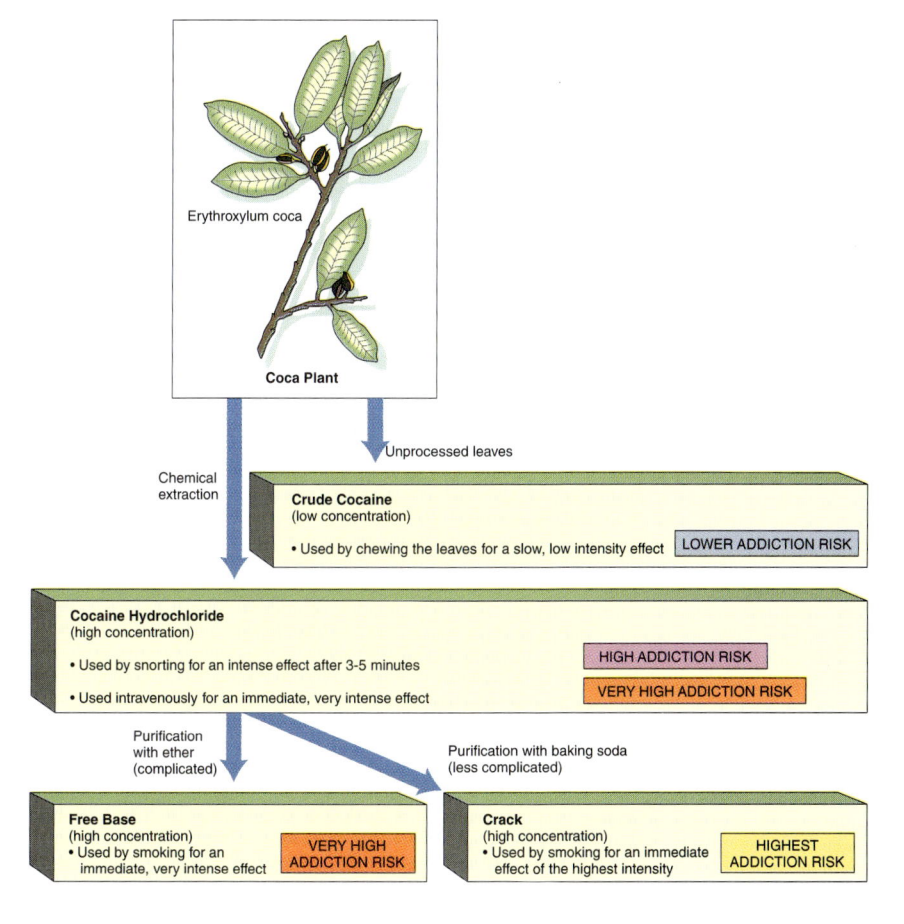

The ways in which the coca plant is processed to make various illegal and dangerous drugs. *(Reproduced by permission of The Gale Group.)*

Cocaine

took to cocaine as a drug they believed had no side effects, was safe, and was not addictive.

Popular overuse

All of these beliefs were eventually seen to be terribly untrue, as a cocaine epidemic in the 1980s claimed many lives, such as that of comedian John Belushi, and wrecked numerous other lives, such as that of the comedian Richard Pryor. Once it is understood what happens to a person's nervous system when he or she ingests or takes in cocaine, it is not surprising that the results are often bad and sometimes tragic. The cocaine sold on the streets is usually a white crystalline powder or an off-white chunky material. It is usually diluted with other substances, like sugar, and is introduced into a person's body by sniffing, swallowing, or injecting it. Most people "snort" the powder or inhale it through their nose, since any of the body's mucous membranes will absorb it into the bloodstream. Injecting the drug means that it must first be turned into a liquid. Both ways create an immediate effect. Smoking "crack" cocaine delivers a more potent high, since crack is distilled cocaine. In its "rock" form it cannot be snorted, but is smoked in pipes. The name "crack" comes from the crackling sound these rock crystals make when heated and burned.

Effects on the brain

However the active part of the drug gets into the body, it delivers the same effect to the person's central nervous system, depending on the amount taken and the user's past drug experience. Usually within seconds, it travels to the brain and produces a sort of overall anesthetic effect because it interferes with the transmission of information from one nerve cell to another. Since this interference is going on within the reward centers of the brain, the user experiences a fairly short-term high that is extremely pleasurable. Physically, the user's heart is racing, and his blood pressure, respiration, and body temperature also increase. The user feels temporarily more alert and energetic. The problem is that these feelings do not last very long, and the user must do more cocaine to recapture them.

In tests with experimental animals, cocaine is the only drug that the animals will repeatedly and continuously demand on their own to the point of killing themselves. Although cocaine is not physically addictive the way heroin is (meaning that the user physically craves the drug and suffers withdrawal when off it), it nonetheless creates a profound psychological dependence in which the mind craves the ecstasy that comes with the drug. Further, since the user experiences fatigue and depression when he or she stops, there is little reason to want to quit. Over time, these cravings get stronger and stronger, and the user can only think of how to get another "hit." This obviously makes them unable to live a normal life without the drug, which has by now taken over their lives.

Effects of abuse

Severe and heavy overuse can make the abuser suffer dizziness, headache, anxiety, insomnia, depression, hallucinations, and have problems moving about. The increase in blood pressure can cause bleeding in the brain as well as breathing problems, both of which have killed many a user. Often, even physically fit people like Len Bias, the All-American basketball star from the University of Maryland, can suddenly die from ingesting cocaine. The medical risks associated with this drug are great, especially since there is no antidote for an overdose. Taking cocaine also has legal consequences, and besides the disorder and dysfunction it brings to a person's life, it can also land them in jail. Many American schools also have a zero-tolerance policy, as do many companies and other organizations.

Overall, despite the glamour that some people see in the drug, the disadvantages far outweigh the temporary advantages, and rather than improving a person's life, it can only do the opposite.

[*See also* **Addiction**]

Cockroaches

Cockroaches are winged insects found in nearly every part of the world. Although they are one of the most primitive living insects, they are very adaptable and highly successful. Some of the species have invaded human habitats and are considered pests since they can spread disease.

"Crazy bug"

Cockroaches or roaches belong to the order Blattaria, which means "to shun the light." They were given this scientific name because they sleep and rest during daylight hours and come out mainly at night. Their common name, however, is a version of the Spanish word *cucaracha*, which means "crazy bug." If you have ever seen one running away from you in a typical wild and zigzagging way, you know how they got their

> **Words to Know**
>
> **Exoskeleton:** An external skeleton.
>
> **Omnivorous:** Plant- and meat-eating.
>
> **Oviparous:** Producing eggs that hatch outside the body.

name. There are some 4,000 species or kinds of cockroaches living in nearly every habitat except Antarctica. All of them prefer to live where it is warm and moist, or where they can at least get water, so it is not surprising that they will move into people's homes if given the chance. Actually, only about 35 of these species are ever associated with people, and the other nearly 4,000 species live throughout the world, although the largest numbers are found in the tropics.

Cockroaches can be interesting, and some would even say fascinating. They can range in length from only 0.1 to 3.2 inches (2.5 to 8.1 centimeters). They seldom use their wings to fly, although some can fly around. Their bodies have a waxy covering that keeps them from drowning. They also can swim and stay underwater for as long as ten minutes. They will rest in one spot without moving for eighteen hours a day, and can go a long time without food. They eat only at night. As for what they eat, they are omnivorous (pronounced om-NIH-vaw-rus), meaning that they can and will eat anything, plant or animal. The more we learn about their diet, the more disgusting they seem, since they eat everything, including animal feces. Although they will eat wood, which is made up of cellulose, they are unable to digest it on their own and, thus, depend on certain protozoa (pronounced pro-toe-ZO-uh) or single-celled organisms that live in their digestive tracts or gut, to break the cellulose down. They make sure they always have these protozoa in their systems by eating the feces of other cockroaches.

A versatile insect

Cockroaches are escape artists whose zigzag darting is done at what seems lightning speed. They can climb easily up vertical surfaces and have such flat bodies that they can hide in the tiniest of cracks and crevices. They have compound eyes (honeycomb-like light sensors) and antennae that are longer than their bodies, which they use to taste, smell, and feel.

They even have a special organ in their mouths that allows them to taste something without actually eating it. Each of their six strong legs has three sets of "knees," all of which can sense vibrations and therefore serve as an early warning system. They also have little motion detectors on their rear end, which explains why they are so hard to catch and stomp. Although females mate only once in their lifetimes, they will stay fertilized all their lives and keep producing eggs without the help of a male cockroach.

Habits and anatomy

Like most other insects, cockroaches have an exoskeleton, meaning their skeleton is located on the outside of their bodies. They have three simple body parts: the head, thorax, and abdomen. Their head is dominated by their long antennae that are constantly moving and sensing the environment. These long, whiplike feelers are used to taste, smell, and feel things, as well as to locate water. To a cockroach, its antennae are more important than its compound eyes on top of its head. Their mouths have jaws that move from side to side instead of up and down, and their versatile mouths allow them to bite, chew, lick, or even lap up their food. They also have unique parts in their mouth called "palpi" that come in handy when humans try to poison them since it allows them to taste something without having actually to eat it.

Its thorax is the middle section of the body; the insect's six legs and two wings are attached to it. Two claws on each foot, plus hairs on their legs, enable them to hold on tightly or climb a wall easily. Their legs are strong and can propel them up to 3 miles (4.8 kilometers) per hour. The abdomen is the largest part of their body, and has several overlapping sections or plates that look like body armor. Their brain is not a single organ in their head, but is rather more like a single nerve that runs the length of their bodies. Their heart is simple, too, looking more like a tube with valves, and their blood is clear. They do not have lungs, but instead breathe through ten pair of holes located on top of the thorax.

Although females mate only once with a male, they stay fertilized and will keep making baby roaches until they die. Most species are oviparous (pronounced o-VIH-puh-rus), meaning the fertilized eggs are laid and hatch outside of

The brown-banded cockroach. *(Reproduced by permission of Edward S. Ross.)*

Coelacanth

the mother's body. She can produce up to fifty babies at once, sometimes within only three weeks. The hatched eggs produce nymphs (pronounced NIMFS), which look like miniature adults. As the new roach grows, it sheds or cracks its outer skin and drops it or molts, growing a new, larger covering. Its does this as much as twelve times before it reaches adulthood. Cockroaches can live anywhere from two to four years.

Aside from most people's natural dislike of any sort of "bug" crawling around where they live, the fact that cockroaches can carry disease-causing microorganisms gives us a very good reason not to want to have them in our homes. Outside or in their natural habitat, they have many natural enemies, including birds, reptiles, mammals, and even other insects. But in our homes none of these usually exist, so cockroaches can reproduce continuously unless removed. Poisons must be used carefully in the home, and it is important first to deny cockroaches access to the indoors by filling cracks to the outdoors. Their food supply can be restricted if we do not leave out any food overnight and keep the kitchen counters and floors swept of crumbs. Leaky faucets and half-full glasses will also provide them with the water they need, so it is important to deny this.

There are four common types of cockroaches that many of us know, sometimes too well. The dark American cockroach is large and is sometimes called a "water bug" or "palmetto bug." The German cockroach is the smallest and has two black streaks down its back. The Australian cockroach is a smaller version of its American cousin, and the Oriental cockroach is reddish-brown or black and is often called a "black beetle." Despite most people's natural dislike of cockroaches, some keep them as pets in an escape-proof terrarium. This recalls an old Italian expression that is translated as "Every cockroach mother thinks her baby is beautiful."

[*See also* **Insects; Invertebrates**]

Coelacanth

A coelacanth (pronounced SEE-luh-kanth) is a large, primitive fish found in the Indian Ocean. Described as a "living fossil" and once thought to be extinct, this deep-sea fish is believed to form one of the "missing links" in the evolution from fish to land animals.

An "extinct" discovery

Until December 1938, the coelacanth was known only by the fossil record that suggested it had lived as long as 350 million years ago in what

> **Words to Know**
>
> **Carnivorous:** Meat-eating.
>
> **Extinct:** No longer alive on Earth.
>
> **Missing link:** An absent member needed to complete a series or resolve a problem.

is called the Devonian period, and that it probably went extinct some 70 million years ago. It was identified scientifically as part of the extinct subclass of *Crossopterygii* (pronounced kross-op-teh-RIH-jee), which means a "lobe-finned fish." Until 1938, most scientists believed the coelacanth had disappeared along with the dinosaurs at the end of the Cretaceous (pronounced kree-TAY-shus) period. However, during that year, fisherman off the eastern coast of South Africa caught a 5-foot (1.5-meter) fish with deep-blue scales and bulging blue eyes that was strange enough to make them bring it to a local museum. The curator, Courtney Latimer, could not identify it, but knew that it was important enough to contact J. L. B. Smith, a leading South African ichthyologist (pronounced ik-thee-OL-low-jist), a zoologist who specializes in fishes. Smith then pronounced the fish to be a coelacanth, and this "living fossil" became the zoological find of the century. Soon after the discovery and publicity, other fisherman from nearby islands were reporting that they too had caught these strange fish that were not good to eat.

Missing link between fish and mammals?

One of the reasons that this discovery caused so much excitement was that in 1938 the coelacanth was thought to be a direct ancestor of tetrapods (pronounced TEH-truh-pods), or four-limbed land animals. This was believable because the coelacanth is unlike any other fish. Coelacanth means "hollow spine" in Greek, and, in fact, this strange creature seems to be a combination of two very different types of fish: those that are made of cartilage, like sharks, and all the other regular bony fishes. Its backbone is a long tube of cartilage instead of being a rigid backbone, yet it has a bony head, teeth, and scales. It is a carnivorous (pronounced kar-NIH-vor-us) predator—meaning that it catches, kills, and eats its live prey—and has impressive jaws and rows of small, sharp teeth. Most

Coelacanth

important, it has four muscular, limblike fins underneath its body that it uses like legs to perch or support itself on the ocean bottom. This led some to believe that it actually used these jointed fins to "walk" on the bottom like a four-legged animal. However, recent molecular analysis indicates that the lungfish, instead of the coelacanth, is genetically the closest living fish that is a relative of land animals.

The modern coelacanth

Since that first discovery of a living coelacanth in 1938, additional coelacanths have been caught not only off the southern tip of Africa but off Sulawesi, Indonesia, as well, suggesting that they are more numerous than believed. Today's coelacanths are larger than those found as fossils, and they can grow to be more than 5 feet (1.5 meters) long and weigh as much as 180 pounds (82 kilograms). Scientists still do not know a great deal about them, and it was not until 1975 when a female was dissected that scientists learned that the coelacanth gives birth to live "pups." Zoologists believe that females do not reach sexual maturity until after 20 years of age and that the gestation (pronounced jes-TAY-shun) period, or the time it takes to develop a newborn, is about 13 months. Females give birth to between 5 and 25 pups, which are capable of surviving on their own after birth.

A preserved specimen of the coelacanth, long thought to be extinct, but discovered living off the coast of Madagascar in the 1980s. It is now on the endangered species list. *(Reproduced by permission of Photo Researchers, Inc.)*

Although there are more coelacanths than at first supposed, they are still recognized as an endangered species. The main reason for this is that they are a highly specialized species that has adapted itself to a narrow habitat range. This means that they can only survive in the cool, deep waters—over 650 feet (200 meters) deep—around volcanic islands. Further, they are a highly specialized fish, resting in lava caves during the day and hunting and feeding at night. Although they move with slow, almost balletlike motion, they are excellent predators who can move surprisingly fast when they ambush a smaller fish for a meal. Along with the nautilus (pronounced NAW-tih-lus) and horseshoe crab, the coelacanth is one of the "living fossils" of the sea, since they have changed little from their ancient ancestors.

[*See also* **Fish**]

Cognition

Cognition is the act of knowing or the process involved in knowing. When we "know" something, it means that we are not only aware or conscious of it, but that we can, in a way, make some sort of judgment about it. Cognition is therefore a very broad term that covers a complicated mental process involving such functions as perception, learning, memory, and problem solving.

How we know

The nature of cognition, or how we know, has been the subject of investigation since the time of the ancient Greeks. It has been studied by both philosophers and scientists. Around 1970, a new field of investigation called cognitive psychology began to emerge. Many of its practitioners study the brain and compare it to a computer in terms of its information storage and retrieval functions. However, most people who study cognition recognize that they are not focusing just on how the brain works as an organ, but are really more concerned with how the *mind* actually works. While there are still several competing theories all trying to explain how the mind works (or how we know), one idea common to most of them is that the mind builds concepts—which are like large symbolic groupings, patterns, or categories—that represent actual things in the real world. It then uses these concepts or patterns that it has already built when it meets a new object or event, and it can then compare the new object to the concept it has already stored.

Cognition

> **Words to Know**
>
> **Cognitive psychology:** School of psychology that focuses on how people perceive, store, and interpret information through such thought processes as memory, language, and problem solving.
>
> **Language:** The use by humans of voice sounds and written symbols representing those sounds in organized combinations to express and to communicate thoughts and feelings.
>
> **Learning:** Thorough knowledge or skill gained by study.
>
> **Memory:** The power or ability of remembering past experiences.
>
> **Perception:** The ability, act, or process of becoming aware of one's surrounding environment through the senses.
>
> **Reasoning:** The drawing of conclusions and judgments through the use of reason.

Elements of cognition

Cognition includes several elements or processes that all work to describe how our knowledge is built up and our judgments are made. Among these many elements are the processes of perceiving, recognizing, conceptualizing, learning, reasoning, problem solving, memory, and language. Some of these processes may include others (for example, problem solving might be considered to be part of reasoning).

Perception. Perception or perceiving refers to the information we get from our five senses (sight, hearing, touch, smell, and taste). Studies have shown that our human senses perceive or take in far more information or data than our nervous systems can ever process or pay attention to. We get around this by organizing this data into chunks or groups, so that when we see a new object (such as a new type of car), we automatically compare it against the vast number of patterns or concepts we already have stored in our brains. When we find that it matches a concept—since we probably already have a general idea of what "carness" is, for example—we do not have to then process every little bit of detailed information about this new car to know that it is a car (that is, in order to perceive it or recognize and understand it as a car). At the end of this process, we have made a judgment of some sort about this new thing. Once scientists discovered this aspect of perception, they were better able to explain how people often see what they expect to see and are sometimes in fact mistaken. This happens when we take only that first, matching impression of something and conclude that it is correct (that is, that the reality is the same as the idea of it we have in our minds) without taking the time to check out all the details of a thing. However, this ability to conceptualize or to create concepts in our minds is very important and is one of the key functions or processes of cognition or knowing.

Reasoning and problem solving. Reasoning could be described as the process by which people systematically develop different arguments and, after consideration, arrive at a conclusion by choosing one. Like reasoning, problem solving also involves comparing things, but it is always aimed at coming to some sort of a solution. We usually do this by creating models of the problem in our minds and then comparing and judging the possible solutions. One thing we know about reasoning and problem solving is that it is usually much more difficult for people to do when it remains in the abstract. In other words, most people can more easily solve a problem if it is concrete than if it remains abstract. A common example given is the game "Rock breaks scissors, scissors cut paper, paper

Cognition

A test of a child's cognition is his or her ability to remember the rules to certain games, and to be able to come up with strategies for winning. *(Reproduced by permission of The Stock Market.)*

Cognition

covers rock." When stated abstractly (A breaks B, B cuts C, and C covers A), we can easily become confused.

Learning. Swiss psychologist Jean Piaget (1896–1980) spent a lifetime studying how children learn, and he identified three stages that children go through as they grow and develop. In the first and simplest stage, an infant believes that an object is still where he or she first saw it, even though the infant had seen it moved to another place. In the second stage, the young child knows that it is at times separate from its environment

Cognition can be demonstrated by children when they find patterns and strategies for success. *(Reproduced by permission of Field Mark Publications.)*

and has developed concepts for things whether he or she is presently involved with them or not. The final, more mature stage has the older child understanding how to use symbols for things (such as things having names) and developing the ability to speak and use those symbols in language.

Memory. Memory, or the ability to recall something that was learned, is another cognitive function that is very important to learning. Scientists usually divide it into short-term and long-term memory. Our short-term memory seems to have a limited capacity, is very much involved with our everyday speech, and appears very important to our identity or our sense of self (who we are). Long-term memory stores information for much longer periods of time and seems to show no limitations at all. The three basic processes common to both types of memory—encoding or putting information into memory, storage, and retrieval—are exactly those used in today's computers.

Language. Although many animals besides human beings have a brain, nervous system, and some cognitive functions (that is, they share in a way many of the same processes of cognition), the one function of cognition that sets humans apart from other animals is the ability to communicate through language. Humans are unique in that they can express concepts as words. Some say that it is through studying language that we will gain an understanding of how the mind works. We do know that we form sentences with our words that allows us to express not just a single concept but complex ideas, rules, and propositions.

Understanding cognition or figuring out the process involved in knowing is something science has only really just begun. However, the combined work of philosophers, psychologists, and other scientists using new technologies for studying the brain may result in the next great scientific breakthrough—the explanation of how the human brain carries out its mental task of knowing.

[*See also* **Brain; Psychology**]

Colloid

Colloids are mixtures whose particles are larger than the size of a molecule but smaller than particles that can be seen with the naked eye. Colloids are one of three major types of mixtures, the other two being solutions and suspensions. The three kinds of mixtures are distinguished by the size of the particles that make them up. The particles in a solution are about the size

Colloid

Types of Colloids

Dispersed Material	Dispersed in Gas	Dispersed in Liquid	Dispersed in Solid
Gas (bubbles)	Not possible	**Foams:** soda pop; whipped cream; beaten egg whites	**Solid foams:** plaster; pumice
Liquid (droplets)	**Fogs:** mist; clouds; hair sprays	**Emulsions:** milk; blood; mayonnaise	butter; cheese
Solid (grains)	**Smokes:** dust; industrial smoke	**Sols and gels:** gelatin; muddy water; jelly; starch solution	**Solid sol:** pearl; colored glass; porcelain; paper

of molecules, approximately 1 nanometer (1 billionth of a meter) in diameter. Those that make up suspensions are larger than 1,000 nanometers. Finally, colloidal particles range in size between 1 and 1,000 nanometers. Colloids are also called colloidal dispersions because the particles of which they are made are dispersed, or spread out, through the mixture.

Types of colloids

Colloids are common in everyday life. Some examples include whipped cream, mayonnaise, milk, butter, gelatin, jelly, muddy water, plaster, colored glass, and paper.

Every colloid consists of two parts: colloidal particles and the dispersing medium. The dispersing medium is the substance in which the colloidal particles are distributed. In muddy water, for example, the colloidal particles are tiny grains of sand, silt, and clay. The dispersing medium is the water in which these particles are suspended.

Colloids can be made from almost any combination of gas, liquid, and solid. The particles of which the colloid is made are called the dispersed material. Any colloid consisting of a solid dispersed in a gas is called a smoke. A liquid dispersed in a gas is referred to as a fog.

Properties of colloids

Each type of mixture has special properties by which it can be identified. For example, a suspension always settles out after a certain period

Colloid

of time. That is, the particles that make up the suspension separate from the medium in which they are suspended and fall to the bottom of a container. In contrast, colloidal particles typically do not settle out. Like the particles in a solution, they remain in suspension within the medium that contains them.

Colloids also exhibit Brownian movement. Brownian movement is the random zigzag motion of particles that can be seen under a microscope. The motion is caused by the collision of molecules with colloid particles in the dispersing medium. In addition, colloids display the Tyndall effect. When a strong light is shone through a colloidal dispersion, the light beam becomes visible, like a column of light. A common example of this effect can be seen when a spotlight is turned on during a foggy night. You can see the spotlight beam because of the fuzzy trace it makes in the fog (a colloid).

Light shining through a solution of sodium hydroxide (left) and a colloidal mixture. The size of colloidal particles makes the mixture, which is neither a solution nor a suspension, appear cloudy. *(Reproduced by permission of Photo Researchers, Inc.)*

Color

Color is a property of light that depends on the frequency of light waves. Frequency is defined as the number of wave segments that pass a given point every second. In most cases, when people talk about light, they are referring to white light. The best example of white light is ordinary sunlight: light that comes from the Sun.

Light is a form of electromagnetic radiation: a form of energy carried by waves. The term "electromagnetic radiation" refers to a vast range of energy waves, including gamma rays, X rays, ultraviolet rays, visible light, infrared radiation, microwaves, radar, and radio waves. Of all these forms, only one can be detected by the human eye: visible light.

White light and color

White light (such as sunlight) and colors are closely related. A piece of glass or crystal can cause a beam of sunlight to break up into a rainbow: a beautiful separation of colors. The technical term for a rainbow is a spectrum. The colors in a spectrum range from deep purple to brilliant red. One way to remember the colors of the spectrum is with the mnemonic device (memory clue) ROY G. BIV, which stands for Red, Orange, Yellow, Green, Blue, Indigo, and Violet.

English physicist Isaac Newton (1642–1727) was the first person to study the connection between white light and colors. Newton caused a beam of white light to fall on a glass prism and found that the white light was broken up into a spectrum. He then placed a second prism in front of the first and found that the colors could be brought back together into a beam of white light. A rainbow is a naturally occurring illustration of Newton's experiment. Instead of a glass prism, though, it is tiny droplets of rainwater that cause sunlight to break up into a spectrum of colors, a spectrum we call a rainbow.

Color and wavelength

The word "color" actually refers to the light of a particular color, such as red light, yellow light, or blue light. The color of a light beam depends on just one factor: the wavelength of the light. Wavelength is defined as the distance between two exactly identical parts of a given wave. Red light consists of light waves with a wavelength of about 700 nanometers (billionths of a meter), yellow light has wavelengths of about 550 nanometers, and blue light has wavelengths of about 450 nanometers. But the wavelengths of colored light are not limited to specific ranges. For example,

Words to Know

Color: A property of light determined by its wavelength.

Colorant: A chemical substance—such as ink, paint, crayons, or chalk—that gives color to materials.

Complementary colors: Two colors that, when mixed with each other, produce white light.

Electromagnetic radiation: A form of energy carried by waves.

Frequency: The number of segments in a wave that pass a given point every second.

Gray: A color produced by mixing white and black.

Hue: The name given to a color on the basis of its frequency.

Light: A form of energy that travels in waves.

Nanometer: A unit of length; this measurement is equal to one-billionth of a meter.

Pigment: A substance that displays a color because of the wavelengths of light that it reflects.

Primary colors: Colors that, when mixed with each other, produce white light.

Shade: The color produced by mixing a color with black.

Spectrum: The band of colors that forms when white light is passed through a prism.

Tint: The color formed by mixing a given color with white.

Tone: The color formed by mixing a given color with gray (black and white).

Wavelength: The distance between two exactly identical parts of a wave.

waves that have wavelengths of 600, 625, 650, and 675 nanometers would have orange, orangish-red, reddish-orange, and, finally, red colors.

The color of objects

Light can be seen only when it reflects off some object. For example, as you look out across a field, you cannot see beams of light passing

Color

through the air, but you *can* see the green of trees, the brown of fences, and the yellow petals of flowers because of light reflected by these objects.

To understand how objects produce color, imagine an object that reflects all wavelengths of light equally. When white light shines on that object, all parts of the spectrum are reflected equally. The color of the object is white. (White is generally not regarded as a color but as a combination of all colors mixed together.)

Now imagine that an object absorbs (soaks up) all wavelengths of light that strike it. That is, no parts of the spectrum are reflected. This object is black, a word that is used to describe an object that reflects no radiation.

Finally, imagine an object that reflects light with a wavelength of about 500 nanometers. Such an object will absorb all wavelengths of light except those close to 500 nanometers. It will be impossible to see red light (700 nanometers), violet light (400 nanometers), or blue light (450 nanometers) because those parts of the spectrum are all absorbed by the object. The only light that is reflected—and the only color that can be seen—is green, which has a wavelength of about 500 nanometers.

Primary and complementary colors

White light can be produced by combining all colors of the spectrum at once, as Newton discovered. However, it is also possible to make white light by combining only three colors in the spectrum: red, green, and blue. For this reason, these three colors of light are known as the primary colors. (For more on the concept of primary colors, see subhead titled "Pigments.") In addition to white light, all colors of the spectrum can be produced by an appropriate mixing of the primary colors. For example, red and green lights will combine to form yellow light.

It is also possible to make white light by combining only two colors, although these two colors are not primary colors. For example, the combination of a bluish-violet light and a yellow light form white light. Any two colors that produce white light, such as bluish-violet and yellow, are known as complementary colors.

The language of colors

A special vocabulary is used to describe colors. The fundamental terms include:

Hue: The basic name of a color, as determined by its frequency. Light with a wavelength of 600 nanometers is said to have an orange hue.

Gray: The color produced by mixing white and black.

Shade: The color produced by mixing a color with black. For example, the shade known as maroon is formed by mixing red and black.

Tint: The color formed by mixing a color with white. Pink is produced when red and white are mixed.

Tone: The color formed by mixing a color with gray (black and white). Red plus white plus black results in the tone known as rose.

Pigments

A pigment is a substance that reflects only certain wavelengths of light. Strictly speaking, there is no such thing as a white pigment because such a substance would reflect all wavelengths of light. A red pigment is one that reflects light with a wavelength of about 700 nanometers; a blue pigment is one that reflects light with a wavelength of about 450 nanometers.

The rules for combining pigment colors are different from those for combining light colors. For example, combining yellow paint and blue paint produces green paint. Combining red paint with yellow paint produces orange paint. And combining all three of the primary colors of paints—yellow, blue, and red—produces black paint.

An array of bright colors. *(Reproduced by permission of The Stock Market.)*

Other color phenomena

Color effects occur in many different situations in the natural world. For example, the swirling colors in a soap bubble are produced by interference, a process in which light is reflected from two different surfaces very close to each other. The soap bubble is made of a very thin layer of soap: the inside and outside surfaces are less than a millimeter away from each other. When light strikes the bubble, then, it is reflected from both the outer surface and from the inside surface of the bubble. The two reflected beams of light interfere with each other in such a way that some wavelengths of light are reinforced, while others are canceled out. It is by this mechanism that the colors of the soap bubble are produced.

[*See also* **Light; Spectroscopy**]

Combustion

Combustion is the chemical term for a process known more commonly as burning. It is one of the earliest chemical changes noted by humans, due at least in part to the dramatic effects it has on materials. Early humans were probably amazed and frightened by the devastation resulting from huge forest fires or by the horror of seeing their homes catch fire and burn. But fire (combustion)—when controlled and used correctly—was equally important to their survival, providing a way to keep warm and to cook their meals.

Today, the mechanism by which combustion takes place is well understood and is more correctly defined as a form of oxidation. This oxidation occurs so rapidly that noticeable heat and light are produced. In general, the term "oxidation" refers to any chemical reaction in which a substance reacts with oxygen. For example, when iron is exposed to air, it combines with oxygen in the air. That form of oxidation is known as rust. Combustion differs from rust in that the oxidation occurs much more rapidly, giving off heat in the process.

History

Probably the earliest scientific attempt to explain combustion was made by Johann Baptista van Helmont, a Flemish physician and alchemist who lived from 1580 to 1644. Van Helmont observed the relationship between a burning material and the resulting smoke and flame it produced. He concluded that combustion involved the escape of a "wild spirit"

> ## Words to Know
>
> **Chemical bond:** Any force of attraction between two atoms.
>
> **Fossil fuel:** A fuel that originates from the decay of plant or animal life; coal, oil, and natural gas are the fossil fuels.
>
> **Industrial Revolution:** The period, beginning about the middle of the eighteenth century, during which humans began to use steam engines as a major source of power.
>
> **Internal-combustion engine:** An engine in which the chemical reaction that supplies energy to the engine takes place within the walls of the engine (usually a cylinder) itself.
>
> **Oxide:** An inorganic compound (one that does not contain carbon) whose only negative part is the element oxygen.
>
> **Thermochemistry:** The science that deals with the quantity and nature of heat changes that take place during chemical reactions and/or changes of state (for instance, from solid to liquid or gas).

(*spiritus silvestre*) from the burning material. This explanation was later incorporated into the phlogiston theory (pronounced flow-JIS-ten), a way of viewing combustion that dominated the thinking of scholars for the better part of two centuries.

According to the phlogiston theory, combustible materials contain a substance—phlogiston—that is given off by the material as it burns. A noncombustible material, such as ashes, will not burn, according to this theory, because all phlogiston contained in the original material (such as wood) had been driven out. The phlogiston theory was developed primarily by German alchemist Johann Becher (1635–1682) and his student Georg Ernst Stahl (1660–1734) at the end of the seventeenth century.

Although scoffed at today, the phlogiston theory explained what was known about combustion at the time of Becher and Stahl. One serious problem with the theory, however, involved weight changes. Many objects actually weigh more after being burned than before. How this could happen when phlogiston escaped from the burning material? One explanation that was offered was that phlogiston had negative weight. Many early chemists thought that such an idea was absurd, but others were willing to consider the possibility. In any case, precise measurements had not

Combustion

yet become an important feature of chemical studies, so loss of weight was not a huge barrier to the acceptance of the phlogiston concept.

Modern theory

Even with all its problems, the phlogiston theory remained popular among chemists for many years. In fact, it was not until a century later that someone proposed a radically new view of the phenomenon. That person was French chemist Antoine Laurent Lavoisier (1743–1794). One key hint

A sulfur combustion. (Reproduced by permission of Photo Researchers, Inc.)

that helped unravel the mystery of the combustion process was the discovery of oxygen by Swedish chemist Karl Wilhelm Scheele (1742–1786) in 1771 and by English chemist Joseph Priestley (1733–1804) in 1774.

Lavoisier used this discovery to frame a new definition of combustion. Combustion, he theorized, is the process by which some material combines with oxygen. For example, when coal burns, carbon in the coal combines with oxygen to form carbon dioxide. Proposing a new theory of combustion was not easy. But Lavoisier conducted a number of experiments involving very careful weight measurements. His results were so convincing that the new oxidation theory was widely accepted in a relatively short period of time.

Lavoisier began another important line of research related to combustion. This research involved measuring the amount of heat generated during oxidation. His earliest experiments involved the study of heat lost by a guinea pig during respiration (breathing), which Lavoisier called a combustion. He was assisted in his work by another famous French scientist, Pierre Simon Laplace (1749–1827).

As a result of their research, Lavoisier and Laplace laid down one of the fundamental principles of thermochemistry, the study of heat changes that take place during chemical reactions. The duo found that the amount of heat needed to decompose (break down) a compound is the same as the amount of heat liberated (freed, or given up) during the compound's formation from its elements. This line of research was further developed by Swiss-Russian chemist Henri Hess (1802–1850) in the 1830s. Hess's development and extension of the work of Lavoisier and Laplace has earned him the title of father of thermochemistry.

Heat of combustion

From a chemical standpoint, combustion is a process in which some chemical bonds are broken and new chemical bonds are formed. The net result of these changes is a release of energy, known as the heat of combustion. For example, suppose that a gram of coal is burned in pure oxygen with the formation of carbon dioxide as the only product. The first step in this reaction requires the breaking of chemical bonds between carbon atoms and between oxygen atoms. In order for this step to occur, energy must be added to the coal/oxygen mixture. For example, a lighted match must be touched to the coal.

Once the carbon-carbon and oxygen-oxygen bonds have been broken, new bonds can be formed. These bonds join carbon atoms with oxygen atoms in the formation of carbon dioxide. The carbon-oxygen bonds contain less energy than did the original carbon-carbon and oxygen-oxygen bonds. The excess energy is released in the form of heat—the heat of combustion. The heat of combustion of one mole of carbon, for example, is about 94 kilocalories. That number means that each time one mole of carbon is burned in oxygen, 94 kilocalories of heat are given off. (A mole is a unit used to represent a certain number of particles, usually atoms or molecules.)

Applications

Humans have been making practical use of combustion for thousands of years. Cooking food and heating homes have long been two major applications of the combustion reaction. With the development of the steam engine by Denis Papin, Thomas Savery, Thomas Newcomen, and others at the beginning of the eighteenth century, however, a new use for combustion was found: performing work. Those first engines employed the combustion of some material, usually coal, to produce heat that was used to boil water. The steam that was produced was then able to move pistons (sliding valves) and drive machinery. That concept is essentially the same one used today to operate fossil-fueled electrical power plants.

Before long, inventors found ways to use steam engines in transportation, especially in railroad engines and steam ships. However, it was not until the discovery of a new type of fuel—gasoline and its chemical relatives—and a new type of engine—the internal-combustion engine—that modern methods of transportation became common. Today, most forms of transportation depend on the combustion of a hydrocarbon fuel (a compound of hydrogen and carbon) such as gasoline, kerosene, or diesel oil to produce the energy that drives pistons and moves vehicles.

Environmental issues

The use of combustion as a power source has had such a dramatic influence on human society that the period after 1750 has sometimes been called the Fossil Fuel Age. Still, the widespread use of combustion for human applications has always caused significant environmental problems. Pictures of the English countryside during the Industrial Revolution (a major change in the economy that resulted from the introduction of power-driven machinery in the mid-eighteenth century), for example, usually show huge clouds of smoke given off by the burning of wood and coal in steam engines.

At the dawn of the twenty-first century, modern societies continued to face environmental problems created by the enormous combustion of carbon-based fuels. For example, one product of any combustion reaction in the real world is carbon monoxide. Carbon monoxide is a toxic (poisonous; potentially deadly) gas that sometimes reaches dangerous concentrations in urban areas around the world. Oxides of sulfur (produced by the combustion of impurities in fuels) and oxides of nitrogen (produced at high temperatures) can also have harmful effects. The most common problem associated with these oxides is the formation of acid rain and smog. Even carbon dioxide itself, the primary product of combustion, can be a problem: it is thought to be at the root of recent global climate changes because of the enormous concentrations it has reached in the atmosphere.

[*See also* **Chemical bond; Heat; Internal-combustion engine; Oxidation-reduction reaction; Pollution**]

Comet

A comet—a Greek word meaning "long-haired"—is best described as a dirty snowball. It is a clump of rocky material, dust, and frozen methane, ammonia, and water that streaks across the sky on a long, elliptical (oval-shaped) orbit around the Sun. A comet consists of a dark, solid nucleus (core) surrounded by a gigantic, glowing mass (coma). Together, the core and coma make up the comet's head, seen as a glowing ball from which streams a long, luminous tail. The tail (which always points away from the Sun) is formed when a comet nears the Sun and melted particles and gases from the comet are swept back by the solar wind (electrically charged particles that flow out from the Sun). A tail can extend as much as 100 million miles (160 million kilometers) in length.

Age-old fascination

Through the ages, comets were commonly viewed as omens, both good and bad, because of their unusual shape and sudden appearance. A comet appearing in 44 B.C. shortly after Roman dictator Julius Caesar was murdered was thought to be his soul returning. A comet that appeared in 684 was blamed for an outbreak of the plague that killed thousands of people.

For centuries, many people believed Earth was at the center of the solar system, with the Sun and other planets orbiting around it. They also believed that comets were a part of Earth's atmosphere. In the sixteenth century, Polish astronomer Nicolaus Copernicus (1473–1543) proposed a theory that placed the Sun at the center of the solar system, with Earth

Comet

> ### Words to Know
>
> **Astronomical unit (AU):** Standard measure of distance to celestial objects, equal to the average distance from Earth to the Sun: 93 million miles (150 million kilometers).
>
> **Coma:** Glowing cloud of mass surrounding the nucleus of a comet.
>
> **Ellipse:** An oval or elongated circle.
>
> **Interstellar medium:** Space between stars, consisting mainly of empty space with a very small concentration of gas atoms and tiny solid particles.
>
> **Nucleus:** Core or center of a comet.

and the other planets in orbit around it. Once astronomers finally determined that comets existed in space beyond Earth's atmosphere, they tried to determine the origin, formation, movement, shape of orbit, and meaning of comets.

Halley's comet

In 1687, English astronomer Edmond Halley (pronounced HAL-ee; 1656–1742) calculated the paths traveled by 24 comets. Among these, he found three—those of 1531, 1607, and 1682—with nearly identical paths. This discovery led him to conclude that comets follow an orbit around the Sun, and thus reappear periodically. Halley predicted that this same comet would return in 1758. Although he did not live to see it, his prediction was correct, and the comet was named Halley's comet. Usually appearing every 76 years, the comet passed by Earth in 1835, 1910, and 1986.

During its last pass over the planet, Halley's comet was explored by the European Space Agency probe *Giotto*. The probe came within 370 miles (596 kilometers) of Halley's center, capturing fascinating images of the 9-mile-long, 5-mile-wide (15-kilometer-long, 8-kilometer-wide) potato-shaped core marked by hills and valleys. Two bright jets of dust and gas, each 9 miles (15 kilometers) long, shot out of the core. *Giotto*'s instruments detected the presence of water, carbon, nitrogen, and sulfur molecules. It also found that the comet was losing about 30 tons of water and 5 tons of dust each hour. This means that although the comet will survive for hundreds more orbits, it will eventually disintegrate. Halley's comet will next pass by Earth in the year 2061.

Comet Hale-Bopp

On July 22, 1995, American astronomer Alan Hale and American amateur stargazer Thomas Bopp independently discovered a new comet just beyond the orbit of Jupiter. Considered by many astronomers to be one of the greatest comets of all time, Comet Hale-Bopp is immense. Its core is almost 25 miles (40 kilometers) in diameter, more than 10 times that of the average comet and 4 times that of Halley's comet. Hale-Bopp's closest pass to Earth occurred on March 22, 1997, when it was 122 million miles (196 million kilometers) away. Despite its great distance from Earth, the huge comet was visible to the naked eye for months before and after that date. Astronomers believed it was one of the longest times any comet had been visible. They estimate that Hale-Bopp will next visit the vicinity of Earth 3,000 years from now.

Comet West above Table Mountain in California shortly before sunrise in March 1976. The comet's bright head is seen just above the mountains, while its broad dust tail sweeps up and back. *(Reproduced courtesy of National Aeronautics and Space Administration.)*

Nourishing snowballs

In mid-1997, scientists announced that small comets about 40 feet (12 meters) in diameter are entering Earth's atmosphere at a rate of about 43,000 a day. The discovery was made by the polar satellite launched by the National Aeronautics and Space Administration (NASA) in early 1996. American physicist Louis A. Frank, the principal scientist for the visible imaging system of the satellite, first proposed the existence of the bombarding comets in 1986.

These comets do not strike the surface of Earth because they break up at heights of 600 to 15,000 miles (960 to 24,000 kilometers) above ground. Sunlight then vaporizes the remaining small icy fragments into huge clouds. As winds disperse these clouds and they sink lower in the atmosphere, the water vapor contained within condenses and falls to the surface as rain. Scientists estimate that this cosmic rain adds one inch of water to Earth's surface every 10,000 to 20,000 years. Over the immense span of Earth's history (4.5 billion years), this amount of water could have been enough to fill the oceans.

Scientists also speculate that the simple organic chemicals (carbon-rich molecules) these comets contain might have fallen on Earth as it was first developing. They may have provided the groundwork for the development of the wide range of life on the planet.

The origin of comets

Comets are considered among the most primitive bodies in the solar system. They are probably debris from the formation of our sun and planets some 4.5 billion years ago. The most commonly accepted theory about where comets originate was suggested by Dutch astronomer Jan Oort in 1950. He believed that over 100 billion inactive comets lie at the frigid, outer edge of the solar system, somewhere between 50,000 and 150,000 astronomical units (AU) from the Sun. (One AU equals the distance from Earth to the Sun.) They remain there in an immense band, called the Oort cloud, until the gravity of a passing star jolts a comet into orbit around the Sun.

In 1951, another Dutch astronomer, Gerard Kuiper, suggested that there is a second reservoir of comets located just beyond the edge of our solar system, around 1,000 times closer to the Sun than the Oort cloud. His hypothetical Kuiper Belt was confirmed in 1992 when astronomers discovered the first small, icy object in a ring of icy debris orbiting the Sun. This ring is located between Neptune and Pluto (sometimes beyond Pluto, depending on its orbit), some 3.6 billion miles (5.8 billion kilometers) from Earth. Since 1992, astronomers have discovered more than 150 Kuiper Belt objects. Many of them are upwards of 60 miles (96 kilometers) in diameter. Several are much larger. In 2000, astronomers discovered one, which they call Varuna, that measures 560 miles (900 kilometers) in diameter, about one-third the size of the planet Pluto. Astronomers believe the ring is filled with hundreds of thousands of small, icy objects that are well-preserved remnants of the early solar system. They are interested in studying these objects because they want to know more about how Earth and the other major planets formed.

The death of comets

There are many theories as to what happens at the end of a comet's life. The most common is that the comet's nucleus splits or explodes, which may produce a meteor shower. It has also been proposed that comets eventually become inactive and end up as asteroids. One more theory states that gravity or some other disturbance causes a comet to exit the solar system and travel out into the interstellar medium.

[*See also* **Meteors and Meteorites**]

Compact disc

A compact disc (CD), or optical disc, is a thin, circular wafer of clear plastic and metal measuring 4.75 inches (120 centimeters) in diameter with a small hole in its center. CDs store different kinds of data or information: sound, text, or pictures (both still and moving). Computer data is stored on CDs in a format called CD-ROM (Compact Disc-Read Only Memory).

All CDs and CD-ROMs are produced the same way. Digital data (the binary language of ones and zeroes common to all computers) is encoded onto a master disc, which is then used to create copies of itself. A laser burns small holes, or pits, into a microscopic layer of metal, usually aluminum. These pits correspond to the binary ones. Smooth areas of the disc untouched by the laser, called land, correspond to the binary zeros. After the laser has completed burning all the pits, the metal is coated with a protective layer.

Audio CDs

Audio or music CDs were introduced in 1982. They offered many advantages over phonograph records and audio tapes, including smaller

Compact disc

size and better sound quality. By 1991, CDs had come to dominate the record industry. In an audio CD player, a small infrared laser shines upon the pits and land on the metal layer of the disc as the disc spins. Land reflects the laser light while pits do not. A mirror or prism between the laser and the disc picks up the reflected light and bounces it onto a photosensitive diode (an electronic device that is sensitive to light). The diode converts the light into a coded string of electrical impulses. The impulses are then transformed into waves for playback through stereo speakers.

CD-ROMs

CD-ROMs (and audio CDs) contain information that cannot be erased or added to once the discs have been created. While audio CDs contain only sound information, CD-ROMs store incredible amounts of

A compact disc.
(Reproduced by permission of Kelly A. Quin.)

text, graphic (video), or sound information. Discs that contain information in more than one of these media are referred to as multimedia. Since video and sound require large amounts of disc storage space, most multimedia CD-ROMs are text-based with some video or sound features added. Information on a CD-ROM is retrieved the same way it is on an audio CD: a laser beam scans tracks of microscopic holes on a rotating disc, eventually converting the information into the proper medium. Because of their high information storage capacity, CD-ROMs have become the standard format for such large published works as software documentation and encyclopedias.

WORMs

WORM (Write Once, Read Many) systems are a little more complicated than CD-ROM systems. Writable WORM discs are made of different material than consumer CD-ROMs. When a WORM disc is created, a laser does not burn pits into a microscopic layer of metal as with a CD-ROM. Instead, in a heat-sensitive film a laser creates distortions that reflect light. These distortions represent bits of data. To read the disc, the laser is scanned over the surface at lower power. A detector then reads and decodes the distortions to obtain the original signal.

WORM discs allow the user to write new information onto the optical disc. Multiple writing sessions may be needed to fill the disc. Once recorded, however, the data is permanent. It cannot be rewritten or erased. WORM discs are especially suited to huge databases (like those used by banks, insurance companies, and government offices) where information might expand but not change.

MODs

Magneto-optical discs (MODs) are rewritable, and operate differently than either ROM or WORM disc. Data is not recorded as distortions of a heat-sensitive layer within the disc. Rather, it is written using combined magnetic and optical techniques. Digital data (binary ones and zeros) is encoded in the optical signal from the laser in the usual manner. Unlike the ROM or WORM discs, however, the MOD write layer is magnetically sensitive. An external magnet located on the write/read head aligns the binary ones and zeros in different directions. The MOD is read by scanning a laser over the spinning disc and evaluating the different directions of the digital data. The MOD is erased by orienting the external magnet so that digital zeros are recorded over the whole disc.

[*See also* **Computer, digital; DVD technology; Laser**]

Complex numbers

Complex numbers are numbers that consist of two parts, one real and one imaginary. An imaginary number is the square root of a real number, such as $\sqrt{-4}$. The expression $\sqrt{-4}$ is said to be imaginary because no real number can satisfy the condition stated. That is, there is no number that can be squared to give the value -4, which is what $\sqrt{-4}$ means. The imaginary number $\sqrt{-1}$ has a special designation in mathematics. It is represented by the letter i.

Complex numbers can be represented as a binomial (a mathematical expression consisting of one term added to or subtracted from another) of the form a + bi. In this binomial, a and b represent real numbers and i = $\sqrt{-1}$. Some examples of complex numbers are $3 - i$, $\frac{1}{2} + 7i$, and $-6 - 2i$.

The two parts of a complex number cannot be combined. Even though the parts are joined by a plus sign, the addition cannot be performed. The expression must be left as an indicated sum.

History

One of the first mathematicians to realize the need for complex numbers was Italian mathematician Girolamo Cardano (1501–1576). Around 1545, Cardano recognized that his method of solving cubic equations often led to solutions containing the square root of negative numbers. Imaginary numbers did not fully become a part of mathematics, however, until they were studied at length by French-English mathematician Abraham De Moivre (1667–1754), a Swiss family of mathematicians named the Bernoullis, Swiss mathematician Leonhard Euler (1707–1783), and others in the eighteenth century.

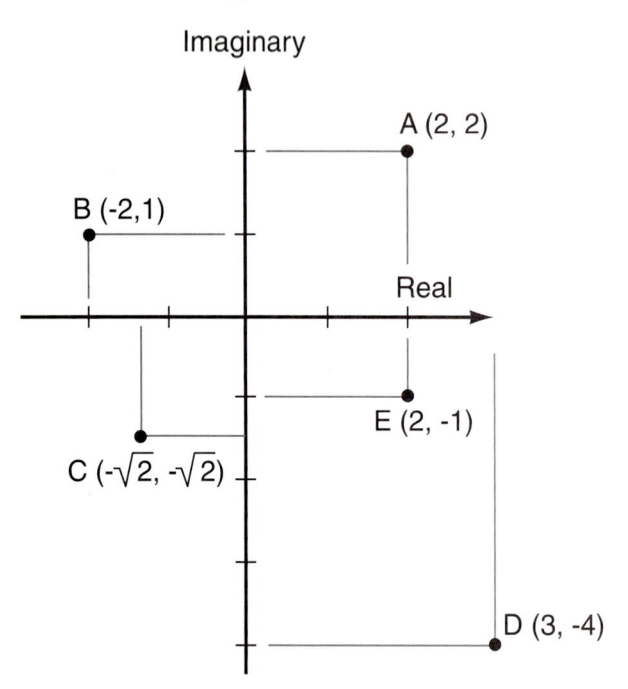

Figure 1. Graphical representation of complex numbers. (Reproduced by permission of The Gale Group.)

Arithmetic

In many ways, operations with complex numbers follow the same rules as those for real numbers. Two exceptions to those rules arise because of the nature of complex numbers. First, what appears to be an addition operation, a + bi, must be left uncombined. Second, the general ex-

534   U·X·L Encyclopedia of Science, 2nd Edition

> **Words to Know**
>
> **Complex number:** A number composed of two separate parts, a real part and an imaginary part, joined by a + sign.
>
> **Imaginary number:** A number whose square (the number multiplied by itself) is a negative number.

pression for any imaginary number, such as $i^2 = -1$, violates the rule that the product of two numbers of a like sign is positive.

The general rules for working with complex numbers are as follows:

1. Equality: To be equal, two complex numbers must have equal real parts and equal imaginary parts. That is, assume that we know that the expressions (a + bi) and (c + di) are equal. That condition can be true if and only if a = c and b = d.

2. Addition: To add two complex numbers, the real parts and the imaginary parts are added separately. The following examples illustrate this rule:

$$(a + bi) + (c + di) = (a + c) + (b + d)i$$
$$(3 + 5i) + (8 - 7i) = 11 - 2i$$

3. Subtraction: To subtract a complex number, subtract the real part from the real part and the imaginary part from the imaginary part. For example:

$$(a + bi) - (c + di) = (a - c) + (b - d)i$$
$$(6 + 4i) - (3 - 2i) = 3 + 6i$$

4. Zero: To equal zero, a complex number must have both its real part and its imaginary part equal to zero. That is, a + bi = 0 if and only if a = 0 *and* b = 0.

5. Opposites: To form the opposite of a complex number, take the opposite of each part. The opposite of a + bi is −(a + bi), or −a + (−b)i. The opposite of 6 − 2i is −6 + 2i.

6. Multiplication: To form the product of two complex numbers, multiply each part of one number by each part of the other. The product of (a + bi) × (c + di) is $ac + adi + bci + bdi^2$. Since $bdi^2 = -bd$, the final product is ac + adi + bci − bd. This expression can be expressed

as a complex number as (ac − bd) + (ad + bc)i. Similarly, the product (5 − 2i) × (4 − 3i) is 14 − 23i.

7. Conjugates: Two numbers whose imaginary parts are opposites are called complex conjugates. The complex numbers a + bi and a − bi are complex conjugates because the terms bi have opposite signs. Pairs of complex conjugates have many applications because the product of two complex conjugates is real. For example, (6 − 12i) × (6 + 12i) = 36 − 144i^2, or 36 + 144 = 180.

8. Division: Division of complex numbers is restricted by the fact that an imaginary number cannot be divided by itself. Division can be carried out, however, if the divisor is first converted to a real number. To make this conversion, the divisor can be multiplied by its complex conjugate.

Graphical representation

After complex numbers were discovered in the eighteenth century, mathematicians searched for ways of representing these combinations of real and imaginary numbers. One suggestion was to represent the numbers graphically, as shown in Figure 1. In graphical systems, the real part of a complex number is plotted along the horizontal axis and the imaginary part is plotted on the vertical axis. Thus, in Figure 1, point A stands for the complex number 2 + 2i and point B stands for the complex number −2 + i.

Uses of complex numbers

For all the "imaginary" component they contain, complex numbers occur frequently in scientific and engineering calculations. Whenever the solution to an equation yields the square root of a negative number (such as $\sqrt{-9}$), complex numbers are involved. One of the problems faced by a scientist or engineer, then, is to figure out what the imaginary and complex numbers represent in the real world.

Composite materials

A composite material (or just composite) is a mixture of two or more materials with properties superior to the materials of which it is made. Many common examples of composite materials can be found in the world around us. Wood and bone are examples of natural composites. Wood

> ## Words to Know
>
> **Fiber:** In terms of composite fillers, a fiber is a filler with one long dimension.
>
> **Matrix:** The part of the composite that binds the filler.
>
> **Particle:** In terms of composite fillers, a particle is a filler with no long dimension.

consists of cellulose fibers embedded in a compound called lignin. The cellulose fibers give wood its ability to bend without breaking, while the lignin makes wood stiff. Bone is a combination of a soft form of protein known as collagen and a strong but brittle mineral called apatite.

Traditional composites

Humans have been using composite materials for centuries, long before they fully understood the structures of such composites. The important building material concrete, for example, is a mixture of rocks, sand, and Portland cement. Concrete is a valuable building material because it is much stronger than any one of the individual components of which it is made. Interestingly enough, two of those components are themselves natural composites. Rock is a mixture of stony materials of various sizes, and sand is a composite of small-grained materials.

Reinforced concrete is a composite developed to further improve the strength of concrete. Steel rods embedded in concrete add both strength and flexibility to the concrete.

Cutting wheels designed for use with very hard materials are also composites. They are made by combining fine particles of tungsten carbide with cobalt powder. Tungsten carbide is one of the hardest materials known, so the composite formed by this method can be used to cut through almost any natural or synthetic material.

Some forms of aluminum siding used in homes are also composite materials. Thin sheets of aluminum metal are attached to polyurethane foam. The polyurethane foam is itself a composite consisting of air mixed with polyurethane. Joining the polyurethane foam to the aluminum makes the aluminum more rigid and provides excellent insulation, an important property for the walls of a house.

Composite materials

In general, composites are developed because no single structural material can be found that has all of the desired characteristics for a given application. Fiber-reinforced composites, for example, were first developed to replace aluminum alloys (mixtures), which provide high strength and fairly high stiffness at low weight but corrode rather easily and can break under stress.

Composite structure

Composites consist of two parts: the reinforcing phase and the binder, or matrix. In reinforced concrete, for example, the steel rods are the reinforcing phase; the concrete in which the rods are embedded are the binder or matrix.

In general, the reinforcing phase can exist in one of three forms: particles, fibers, or flat sheets. In the cutting wheels described above, for example, the reinforcing phase consists of tiny particles of cobalt metal in a binder of tungsten carbide. A plastic fishing rod is an example of a composite in which the reinforcing phase is a fiber. In this case, the fiber is made of threadlike strips of glass placed in an epoxy matrix. (Epoxy is a strong kind of plastic.) An example of a flat sheet reinforcing phase is plywood. Plywood is made by gluing together thin layers of wood so that the wood grain runs in different directions.

The binder or matrix in each of these cases is the material that supports and holds in place the reinforcing material. It is the tungsten carbide in the cutting wheel, the epoxy plastic in the fishing rod, or the glue used to hold the sheets of wood together.

High-performance composites

High-performance composites are composites that perform better than conventional structural materials such as steel and aluminum alloys. They are almost all fiber-reinforced composites with polymer (plasticlike) matrices.

The fibers used in high-performance composites are made of a wide variety of materials, including glass, carbon, boron, silicon carbide, aluminum oxide, and certain types of polymers. These fibers are generally interwoven to form larger filaments or bundles. Thus, if one fiber or a few individual fibers break, the structural unit as a whole—the filament or bundle—remains intact. Fibers usually provide composites with the special properties, such as strength and stiffness, for which they are designed.

In contrast, the purpose of the matrix in a high-performance composite is to hold the fibers together and protect them from damage from the outside environment (such as heat or moisture) and from rough handling. The matrix also transfers the load placed on a composite from one fiber bundle to the next.

Most matrices consist of polymers such as polyesters, epoxy vinyl, and bismaleimide and polyimide resins. The physical properties of any given matrix determine the ultimate uses of the composite itself. For example, if the matrix melts or cracks at a low temperature, the composite can be used for applications only at temperatures less than that melting or cracking point.

Composting

Composting is the process of arranging and manipulating plant and animal materials so that they are gradually broken down, or decomposed, by soil bacteria and other organisms. The resulting decayed organic matter is a black, earthy-smelling, nutritious, spongy mixture called compost or humus. Compost is usually mixed with other soil to improve the

Backyard composting in Livonia, Michigan. This system includes a composting bin made from chicken wire and plywood, a soil screen made from mesh wire and wood boards, a wheelbarrow, and a digging fork. *(Reproduced by permission of Field Mark Publications.)*

Composting

> ### Words to Know
>
> **Decomposition:** The breakdown of complex organic materials into simple substances by the action of microorganisms.
>
> **Humus:** Decayed plant or animal matter.
>
> **Microorganism:** A living organism that can only be seen through a microscope.
>
> **Nutrient:** Any substance required by a plant or animal for energy and growth.
>
> **Organic:** Made of or coming from living matter.

soil's structural quality and to add nutrients for plant growth. Composting is a method used by gardeners to produce natural fertilizer for growing plants.

Why compost?

Compost added to soil aids in its ability to hold oxygen and water and to bind to certain nutrients. It improves the structure of soils that are too sandy to hold water or that contain too much clay to allow oxygen to penetrate. Compost also adds mineral nutrients to soil. Compost mixed with soil makes the soil darker, allowing it to absorb the Sun's heat and warm up faster in the spring.

Adding compost to soil also benefits the environment. The improved ability of the soil to soak up water helps to prevent soil erosion caused by rainwater washing away soil particles. In addition, composting recycles organic materials that might otherwise be sent to landfills.

Composting on any scale

Composting can be done on a small scale by homeowners using a small composting bin or a hole where kitchen wastes are mixed with grass clippings, small branches, shredded newspapers, or other organic matter. Communities may have large composting facilities to which residents bring grass, leaves, and branches to be composted as an alternative to disposal in a landfill. Sometimes sewage sludge, the semisolid material from sewage treatment plants, is added. The resulting humus is used to condition soil on golf courses, parks, and other municipal grounds.

Compound, chemical

Materials to compost

Most organic (carbon-containing) materials can be composted—shredded paper, hair clippings, food scraps, coffee grounds, eggshells, fireplace ashes, chopped-up Christmas trees, and seaweed among them. Meat is omitted because it can give off bad odors during decomposition and attract rats and other pests. The microorganisms needed to break down the organic matter are supplied by adding soil or humus to the compost heap. Manure from farm or zoo animals makes an excellent addition to compost. Wastes from household pets are not used because they may carry disease.

How a compost heap works

A compost heap needs both water and oxygen to work efficiently. More importantly, the contents must be turned regularly to expose all areas to oxygen, which raises the temperature of the compost.

The processes that occur within a compost heap are microbiological, chemical, and physical. Microorganisms break down the chemical bonds of organic materials in the presence of oxygen and moisture, giving off heat. Some organisms work on the compost pile physically after it has cooled to normal air temperature. Organisms such as mites, snails, slugs, beetles, and worms digest the organic materials, adding their nutrient-filled excrement to the humus.

The nutrients

During the composting process, organic material is broken down into mineral nutrients such as nitrogen. Plants absorb nutrients through their roots and use them to make chlorophyll, proteins, and other substances needed for growth. Chlorophyll is the green pigment in plant leaves that captures sunlight for photosynthesis, the process in which plants use light energy to manufacture their own food.

[*See also* **Agrochemicals; Recycling; Waste management**]

Compound, chemical

A chemical compound is a substance composed of two or more elements chemically combined with each other. Compounds are one of three

Compound, chemical

Words to Know

Family: A group of chemical compounds with similar structure and properties.

Functional group: A group of atoms that give a molecule certain distinctive chemical properties.

Mixture: A combination of two or more substances that are not chemically combined with each other and that can exist in any proportion.

Molecule: A particle made by the chemical combination of two or more atoms; the smallest particle of which a compound is made.

Octet rule: A hypothesis that atoms having eight electrons in their outermost energy level tend to be stable and chemically unreactive.

Oxide: An inorganic compound whose only negative part is the element oxygen.

---

general forms of matter. The other two are elements and mixtures. Historically, the distinction between compounds and mixtures was often unclear. Today, however, the two can be distinguished from each other on the basis of three primary criteria.

First, compounds have constant and definite compositions, while mixtures may exist in virtually any proportion. A sample of water always consists of 88.9 percent oxygen and 11.1 percent hydrogen. It makes no difference whether the water comes from Lake Michigan, the Grand River, or a cloud in the sky. Its composition is always the same.

By comparison, a mixture of hydrogen and oxygen gases can have any composition whatsoever. You can make a mixture of 90 percent hydrogen and 10 percent oxygen; 75 percent hydrogen and 25 percent oxygen; 50 percent hydrogen and 50 percent oxygen; or any other combination.

Second, the elements that make up a compound lose their characteristic elemental properties when they become part of the compound. In contrast, the elements that make up a mixture retain those properties. In a mixture of iron and sulfur, for example, black iron granules and yellow sulfur crystals often remain recognizable. Also, the iron can be extracted from the mixture by means of a magnet, or the sulfur can be dissolved out with carbon disulfide. Once the compound called iron(II) sulfide has been formed, however, both iron and sulfur lose those properties. Iron

Opposite Page:
The difference between a mixture and a compound. *(Reproduced by permission of The Gale Group.)*

cannot be extracted from the compound with a magnet, and sulfur cannot be dissolved out of the compound with carbon disulfide.

Third, some visible evidence—usually heat and light—accompanies the formation of a compound. But no observable change takes place in the making of a mixture. A mixture of iron and sulfur can be made simply by stirring the two elements together. But the compound iron(II) sulfide is produced only when the two elements are heated. Then, as they combine with each other, they give off a glow.

History

Prior to the 1800s, the term "compound" did not have a precise meaning: the same word was used to describe both a mixture and a compound. Scientists at that time could not measure the composition of materials very accurately. Only very rough balances were available, so no measurement of weight could be trusted to any great extent.

Thus, suppose that a chemist in 1800 reported the composition of water as 88.9 percent oxygen and 11.1 percent hydrogen, and a second chemist reported 88.6 percent oxygen and 11.4 percent hydrogen. The question, then, was whether different samples of water had different compositions or whether the balances used to measure the components were just inaccurate. In the former case, the term compound would have no meaning, since water's composition would not always be the same. In the latter case, water *could* be thought of as a compound, and the differences in composition reported could be attributed to problems with weighing, not with the composition of water.

This debate raged for many years among chemists. Gradually, balances became more and more accurate, and experimental results became more and more consistent. By about 1800, it had become obvious that something like a compound really did exist. And the most important characteristic of the compound was that its composition was always and everywhere exactly the same.

Formation of compounds

Compounds form when two or more elements combine with each other. When a sodium metal is added to

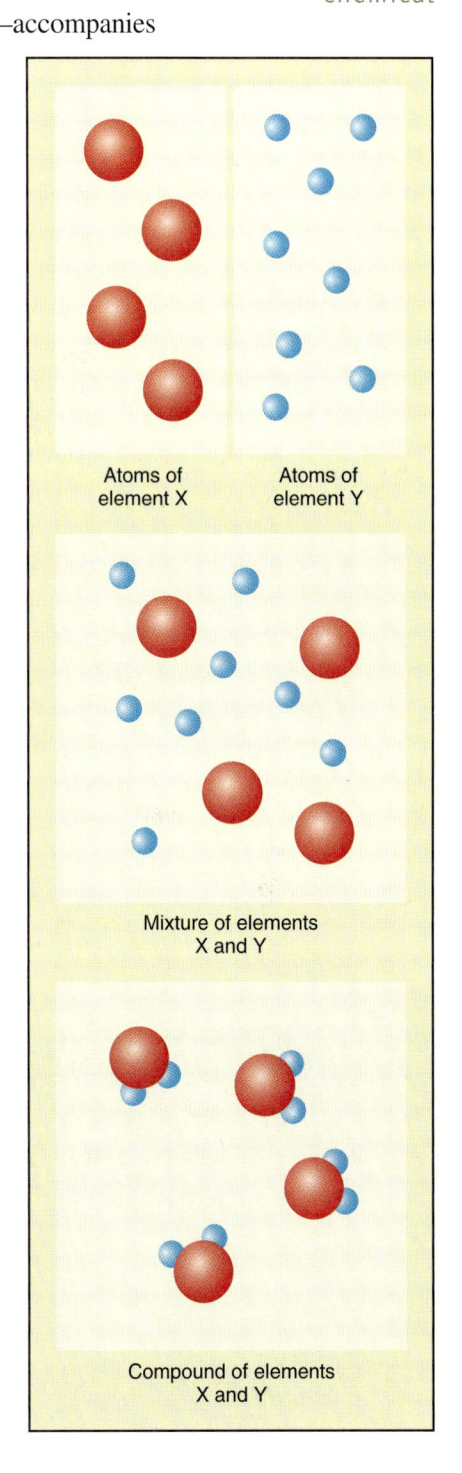

Compound, chemical

Atoms of element X   Atoms of element Y

Mixture of elements X and Y

Compound of elements X and Y

Compound, chemical

> ### Coordination Compounds
>
> What do hemoglobin, chlorophyll, and vitamin B_{12} all have in common? They are members of a class of compounds known as coordination compounds. Coordination compounds consist of two parts: a central metal atom surrounded by a group of atoms known as ligands. Some common ligands are water, ammonia, carbon monoxide, the chloride ion, the cyanide ion, and the thiocyanate ion. In coordination compounds, ligands cluster around the central metal atom in groups of four, six, or some other number. The central difference among hemoglobin, chlorophyll, and vitamin B_{12} is the metal at the center of the compound.
>
> A special class of coordination compounds is known as chelates. Their name comes from the Greek word *chela,* or "claw." The ligands in a chelate grab the metal ion like a claw and hold it tightly. One example of a chelating agent is the compound known as ethylenediaminetetraacetic acid (or EDTA). EDTA can be used to soften water because it "clamps on" to calcium ions that make water hard. It is also used to treat people who have been poisoned by lead, mercury, or other toxic metals. The EDTA grabs onto the metal ions in the blood and removes them from the body, thus preventing harm to a person.

chlorine gas, a burst of light is produced. The elements sodium and chlorine come together to form the compound known as sodium chloride, or ordinary table salt.

The formation of compounds can be understood by examining changes that take place on an atomic level. (An atom is the smallest part of an element that can exist alone.) Those changes are covered by a scientific law known as the octet rule. The octet rule states that all atoms tend to be stable if they have eight electrons (an octet) in their outermost energy level. (That law is modified somewhat for the very lightest elements.) The tendency of elements to combine with each other to form compounds is an effort on the part of atoms to form complete octets. In the case of sodium chloride, a compound is formed when sodium atoms give away electrons to chlorine atoms. At the conclusion of this exchange, both sodium atoms and chlorine atoms have complete outer energy levels.

Atoms can satisfy the octet rule by methods other than the gain and loss of electrons. They can, for example, share electrons with each other. The joining together of two atoms as a result of the gain and loss or sharing of electrons is known as chemical bonding. Chemical bonds are the forces that hold elements together in a compound.

Types of compounds

Most of the ten million or so chemical compounds that are known today can be classified into a relatively small number of subgroups or families. More than 90 percent of these compounds are designated as organic compounds because they contain the element carbon. In turn, organic compounds can be further subdivided into a few dozen major families such as the alkanes, alkenes, alkynes, alcohols, aldehydes, ketones, carboxylic acids, and amines. Each of these families can be recognized by the presence of a characteristic functional group that strongly determines the physical and chemical properties of the compounds that make up that family. For example, the functional group of the alcohols is the hydroxyl group (—OH) and that of the carboxylic acids is the carboxyl group (—COOH).

An important subset of organic compounds are those that occur in living organisms: the biochemical compounds. In general, biochemical compounds can be classified into four major families: carbohydrates, proteins, nucleic acids, and lipids. Members of the first three families are grouped together because of common structural features and similar physical and chemical properties. Members of the lipid family are classified as such on the basis of their solubility (ability to dissolve). They tend not to be soluble in water, but soluble in organic (or carbon-containing) liquids.

Inorganic compounds are typically classified into one of five major groups: acids, bases, salts, oxides, and others. Acids can be defined as compounds that produce hydrogen ions when placed into water. Bases, in contrast, are compounds that produce hydroxide ions when placed into water. Oxides are compounds whose only negative part is oxygen. Salts are compounds that consist of two parts, one positive (the cation) and one negative (the anion). The cation can be of any element or group of elements *except* hydrogen, while the anion may be of any atom or group of atoms *except* the hydroxide group.

This system of classification is useful in grouping compounds that have many similar properties. For example, all acids have a sour taste, leave a pink stain on litmus paper, and react with bases to form salts. One drawback of the system, however, is that it may not give a sense of the enormous diversity of compounds that exist within a particular family. For example, the element chlorine forms at least five common acids, known as hydrochloric, hypochlorous, chlorous, chloric, and perchloric acids. For all their similarities, these five acids also have important distinctive properties.

[*See also* **Element, chemical**]

Computer, analog

A digital computer performs calculations based solely upon numbers or symbols. An analog computer, on the other hand, translates continuously changing quantities (such as temperature, pressure, weight, or speed) into corresponding voltages or gear movements. It then performs "calculations" by comparing, adding, or subtracting voltages or gear motions in various ways. The final result is sent to an output device such as a cathode-ray tube or pen plotter on a roll of paper. Common devices such as thermostats and bathroom scales are actually simple analog computers: they "compute" one thing by measuring another. They do not count.

Early analog computers

The earliest known analog computer is an astrolabe. First built in Greece around the second century B.C., the device uses gears and scales to predict the motions of the Sun, planets, and stars. Other early measuring devices are also analog computers. Sundials trace a shadow's path to show the time of day. The slide rule (a device used for calculation that consists of two rules with scaled numbers) was invented about 1620 and is still used, although it has been almost completely replaced by the electronic calculator.

Modern analog computers

Vannevar Bush, an electrical engineer at the Massachusetts Institute of Technology (MIT), created in the 1930s what is considered to be the first modern computer. He and a team from MIT's electrical engineering staff, discouraged by the time-consuming mathematical computations required to solve certain engineering problems, began work on a device to solve these equations automatically. In 1935, they unveiled the second version of their device, dubbed the "differential analyzer." It weighed 100 tons and contained 150 motors and hundreds of miles of wires connecting relays and vacuum tubes. By present standards the machine was slow, only about 100 times faster than a human operator using a desk calculator.

In the 1950s, RCA produced the first reliable design for a fully electronic analog computer. By this time, however, many of the most complex functions of analog computers were being assumed by faster and more accurate digital computers. Analog computers are still used today for some applications, such as scientific calculation, engineering design, industrial process control, and spacecraft navigation.

[*See also* **Computer, digital**]

Computer, digital

The digital computer is a programmable electronic device that processes numbers and words accurately and at enormous speed. It comes in a variety of shapes and sizes, ranging from the familiar desktop microcomputer to the minicomputer, mainframe, and supercomputer. The supercomputer is the most powerful in this list and is used by organizations such as NASA (National Aeronautics and Space Administration) to process upwards of 100 million instructions per second.

The impact of the digital computer on society has been tremendous; in its various forms, it is used to run everything from spacecraft to factories, health-care systems to telecommunications, banks to household budgets.

The story of how the digital computer evolved is largely the story of an unending search for labor-saving devices. Its roots go back beyond the calculating machines of the 1600s to the pebbles (in Latin, *calculi*) that the merchants of Rome used for counting and to the abacus of the fifth century B.C. Although none of these early devices were automatic, they were useful in a world where mathematical calculations performed by human beings were full of human error.

The Analytical Engine

By the early 1800s, with the Industrial Revolution well underway, errors in mathematical data had grave consequences. Faulty navigational tables, for example, were the cause of frequent shipwrecks. English mathematician Charles Babbage (1791–1871) believed a machine could do mathematical calculations faster and more accurately than humans. In 1822, he produced a small working model of his Difference Engine. The

Computer, digital

machine's arithmetic functioning was limited, but it could compile and print mathematical tables with no more human intervention needed than a hand to turn the handles at the top of the model.

Babbage's next invention, the Analytical Engine, had all the essential parts of the modern computer: an input device, a memory, a central processing unit, and a printer.

Although the Analytical Engine has gone down in history as the prototype of the modern computer, a full-scale version was never built. Even if the Analytical Engine had been built, it would have been pow-

A Bit-Serial Optical Computer (BSOC), the first computer to store and manipulate data and instructions as pulses of light. *(Reproduced by permission of Photo Researchers, Inc.)*

ered by a steam engine, and given its purely mechanical components, its computing speed would not have been great. In the late 1800s, American engineer Herman Hollerith (1860–1929) made use of a new technology—electricity—when he submitted to the United States government a plan for a machine that was eventually used to compute 1890 census data. Hollerith went on to found the company that ultimately became IBM.

Mammoth modern versions

World War II (1939–45) marked the next significant stage in the evolution of the digital computer. Out of it came three mammoth computers. The Colossus was a special-purpose electronic computer built by the British to decipher German codes. The Mark I was a gigantic electromechanical device constructed at Harvard University. The ENIAC was a fully electronic machine, much faster than the Mark I.

The ENIAC operated on some 18,000 vacuum tubes. If its electronic components had been laid side by side two inches apart, they would have covered a football field. The computer could be instructed to change programs, and the programs themselves could even be written to interact with each other. For coding, Hungarian-born American mathematician John von Neumann proposed using the binary numbering system, 0 and 1, rather than the 0 to 9 of the decimal system. Because 0 and 1 correspond to the on or off states of electric current, computer design was greatly simplified.

Since the ENIAC, advances in programming languages and electronics—among them, the transistor, the integrated circuit, and the microprocessor—have brought about computing power in the forms we know it today, ranging from the supercomputer to far more compact personal models.

Future changes to so-called "computer architecture" are directed at ever greater speed. Ultra-high-speed computers may run by using superconducting circuits that operate at extremely cold temperatures. Integrated circuits that house hundreds of thousands of electronic components on one chip may be commonplace on our desktops.

[See also **Computer, analog; Computer software**]

Computer software

Computer software is a package of specific instructions (a program) written in a defined order that tells a computer what to do and how to do it. It is the "brain" that tells the "body," or hardware, of a computer what to

Computer software

> **Words to Know**
>
> **Computer hardware:** The physical equipment used in a computer system.
>
> **Computer program:** Another name for computer software, a series of commands or instructions that a computer can interpret and execute.

do. "Hardware" refers to all the visible components in a computer system: electrical connections, silicon chips, disc drives, monitor, keyboard, printer, etc. Without software, a computer can do nothing; it is only a collection of circuits and metal in a box.

History

The first modern computers were developed by the United States military during World War II (1939–45) to calculate the paths of artillery shells and bombs. These computers used vacuum tubes that had on-off switches. The settings had to be reset by hand for each operation.

These very early computers used the familiar decimal digits (0 to 9) to represent data (information). Computer engineers found it difficult to work with 10 different digits. John von Neumann (1903–1957), a Hungarian-born American mathematician, decided in 1946 to abandon the decimal system in favor of the binary system (a system using only 0 and 1; "bi" means two). That system has been used ever since.

How a computer uses the binary system

Although computers perform seemingly amazing feats, they actually understand only two things: whether an electrical "on" or "off" condition exists in their circuits. The binary numbering system works well in this situation because it uses only the 0 and 1 *bi*nary dig*its* (later shortened to *bits*). Binary 1 represents on and binary 0 represents off. Program instructions are sent to the computer by combining bits together in groups of six or eight. This process takes care of the instructional part of programs.

Computer codes

Data—in the form of decimal numbers, letters, and special characters—also has to be available in the computer. For this purpose, the EBCDIC and ASCII codes were developed.

EBCDIC (pronounced EB-see-dick) stands for Extended Binary Coded Decimal Interchange Code. It was developed by IBM Corporation and is used in most of its computers. In EBCDIC, eight bits are used to represent a single character.

ASCII (pronounced AS-key) is American Standard Code for Information Interchange. ASCII is a seven-bit code developed in a joint effort by several computer manufacturers to create a standard code that could be used on any computer, regardless who made it. ASCII is used in most personal computers today and has been adopted as a standard by the U.S. government.

The development of computer languages

The first modern computer was named ENIAC for Electronic Numerical Integrator And Calculator. It was assembled in 1946. Most programming was done by and for military and scientific users. That began to change after Grace Hopper, an American computer scientist and naval officer, developed FLOW-MATIC, or assembly language, as it came to be called. It uses short names (known as mnemonics or memory aids) to represent common sequences of instructions. These instructions are in turn translated back into the zeroes and ones of machine language when the program is run. This was an important step toward developing "user-friendly" computer software. FLOW-MATIC was one of the first "high-level" computer languages.

Soon, other high-level computer languages were developed. By 1957, IBM had created FORTRAN, a language specifically designed for scientific and engineering work involving complicated mathematical formulas. FORTRAN stands for FORmula TRANslater. It became the first high-level programming language to be used by many computer users. COBOL (COmmon Business Oriented Language) was developed in 1959 to help businesses organize records and manage data files.

During the first half of the 1960s, two professors at Dartmouth College developed BASIC (Beginner's All-purpose Symbolic Instruction Code). This was the first widespread computer language designed for and used by nonprofessional programmers. It was extremely popular throughout the 1970s and 1980s. Its popularity was increased by the development and sale of personal computers, many of which already had BASIC programmed into their memories.

Types of computer software

The development of high-level languages helped to make computers common objects in workplaces and homes. Computers, of course, must

Computer software

have the high-level language command translated back into machine language before they can act on it. The programs needed to translate high-level language back into machine language are called translator programs. They represent another type of computer software.

Operating system software is yet another type of software that must be in a computer before it can read and use commercially available software packages. Before a computer can use application software, such as a word-processing or a game-playing package, the computer must run the instructions through the operating system software. This contains many built-in instructions, so that each piece of application software does not have to repeat simple instructions, like telling the computer how to print something out. DOS (Disc Operating System) is a popular operating system software program for many personal computers used today.

Application software. Once some type of operating system software is loaded into a computer, the computer can load and understand many other types of software. Software can tell computers how to create documents, to solve simple or complex calculations for business people and scientists, to play games, to create images, to maintain and sort files, and to complete hundreds of other tasks.

Word-processing software makes writing, rewriting, editing, correcting, arranging, and rearranging words convenient. Database software enables computer users to organize and retrieve lists, facts, and inventories, each of which may include thousands of items. Graphics software lets you draw and create images.

Desktop publishing software allow people to arrange photos, pictures, and words on a page before any printing is done. With desktop publishing and word-processing software, there is no need for cutting and pasting layouts. Entire books can be written and formatted by the author. The printed copy or even just a computer disk with the file can be delivered to a traditional printer without the need to reenter all the words on a typesetting machine.

Software for games can turn a computer into a spaceship, a battlefield, or an ancient city. As computers get more powerful, computer games get more realistic and sophisticated.

Communications software allows people to send and receive computer files and faxes over phone lines. Transferring files, sending and receiving data, using data stored on another computer, and electronic mail (e-mail) systems that allow people to receive messages in their own "mailboxes" are some common uses of communications software.

The Y2K hubbub

As the end of the 1990s approached, the world became preoccupied or perhaps even obsessed with the coming of the year 2000, nicknamed "Y2K" (Y for year and 2 times K, a standard designation for a thousand). Many feared that at the stroke of midnight between December 31, 1999, and January 1, 2000, computers and computer-assisted devices would come crashing down.

The so-called Y2K bug was a fault built into computer software because early developers of computer programs were uncertain that computers would even have a future. To save on memory and storage wherever possible, these developers built in standardized dates with two digits each for the day, month, and year. For instance, January 2, 1961, was read as 010261. However, this short form could also mean January 2, 1561, or January 2, 2161.

By the mid-1970s, programmers were beginning to recognize the potential obstacle. They began experimenting with plugging 2000-plus

The use of many varied forms of computer software makes the computer an indispensable tool. *(Reproduced by permission of Photo Researchers, Inc.)*

dates into their systems and software; they quickly found the dates did not compute. However, it was not until 1995 that the U.S. Congress, the media, and the public all seemed to "discover" that the end was drawing near. As of 1999, 1.2 trillion lines of computer code needed to be fixed. Left uncorrected, the Y2K bug could have fouled computers that controlled power grids, air traffic, banking systems, and phone networks, among other systems. In response, businesses and governments around the world spent over $200 billion to reprogram and test vulnerable computers.

When the year 2000 became a reality, the anticipated computer glitches never materialized: power plants kept working, airplanes kept flying, and nuclear missiles were kept on the ground. Problems that did arise were minor and were quickly fixed with hardly anyone noticing. There were many other added benefits of the money and time spent on the Y2K problem: Businesses and governments upgraded their computers and other equipment. With the help of the World Bank and other Y2K funders, poorer countries were given machines and Internet connections they were allowed to keep. Many U.S. businesses weeded out older machines, combined similar systems, and catalogued their software and computers. In the end, individuals, businesses, and countries learned to work together to overcome a common problem.

[*See also* **CAD/CAM; Internet**]

Conservation laws

Conservation laws are scientific statements that describe the amount of some quantity before and after a physical or chemical change.

Conservation of mass and energy

One of the first conservation laws to be discovered was the conservation of mass (or matter). Suppose that you combine a very accurately weighed amount of iron and sulfur with each other. The product of that reaction is a compound known as iron(II) sulfide. If you also weigh very accurately the amount of iron(II) sulfide formed in that reaction, you will discover a simple relationship: The weight of the beginning materials (iron plus sulfur) is exactly equal to the weight of the product or products of the reaction (iron(II) sulfide). This statement is one way to express the law of conservation of mass. A more formal definition of the law is that mass (or matter) cannot be created or destroyed in a chemical reaction.

> **Words to Know**
>
> **Angular momentum:** For objects in rotational (or spinning) motion, the product of the object's mass, its speed, and its distance from the axis of rotation.
>
> **Conserved quantities:** Physical quantities, the amounts of which remain constant before, during, and after some physical or chemical process.
>
> **Linear momentum:** The product of an object's mass and its velocity.
>
> **Mass:** A measure of the quantity of matter.
>
> **Subatomic particle:** A particle smaller than an atom, such as a proton, neutron, or electron.
>
> **Velocity:** The rate at which the position of an object changes with time, including both the speed and the direction.

A similar law exists for energy. When you turn on an electric heater, electrical energy is converted to heat energy. If you measure the amount of electricity supplied to the heater and the amount of heat produced by the heater, you will find the amounts are equal. In other words, energy is conserved in the heater. It may take various forms, such as electrical energy, heat, magnetism, or kinetic energy (the energy of an object due to its motion), but the relationship is always the same: The amount of energy used to initiate a change is the same as the amount of energy detected at the end of the change. In other words, energy cannot be created or destroyed in a physical or chemical change. This statement summarizes the law of conservation of energy.

At one time, scientists thought that the law of conservation of mass and the law of conservation of energy were two distinct laws. In the early part of the twentieth century, however, German-born American physicist Albert Einstein (1879–1955) demonstrated that matter and energy are two forms of the same thing. He showed that matter can change into energy and that energy can change into matter. Einstein's discovery required a restatement of the laws of conservation of mass and energy. In some instances, a tiny bit of matter can be created or destroyed in a change. The quantity is too small to be measured by ordinary balances, but it still amounts to something. Similarly, a small amount of energy can be cre-

Conservation laws

ated or destroyed in a change. But, the *total* amount of matter PLUS energy before and after a change still remains constant. This statement is now accepted as the law of conservation of mass and energy.

Examples of the law of conservation of mass and energy are common in everyday life. The manufacturer of an electric heater can tell consumers how much heat will be produced by a given model of heater. The amount of heat produced is determined by the amount of electrical current that goes into the heater. Similarly, the amount of gasoline that can be formed in the breakdown of petroleum can be calculated by the amount of petroleum used in the process. And the amount of nuclear energy produced by a nuclear power plant can be calculated by the amount of uranium-235 used in the plant.

Calculations such as these are never quite as simple as they sound. We think of an electric lightbulb, for example, as a way of changing electrical energy into light. Yet, more than 90 percent of that electricity is actually converted to heat. (Baby chicks are kept warm by the heat of lightbulbs.) Still, the conservation law holds true. The total amount of energy produced in a lightbulb (heat plus light) is equal to the total amount of energy put into the bulb in the form of electricity.

Other conservation laws

Conservation of electric charge. Most physical properties abide by conservation laws. Electric charge is another example. Electric charge is the property that makes you experience a shock or spark when you touch a metal doorknob after shuffling your feet across a rug. It is also the property that produces lightning. Electric charge comes in two varieties: positive and negative.

The law of conservation of electric charge states that the total electric charge in a system is the same before and after any kind of change. Imagine a large cloud of gas with 1,000 positive (+) charges and 950 negative (−) charges. The total electrical charge on the gas would be 1,000+ + 950− = 50+. Next, imagine that the gas is pushed together into a much smaller volume. Whatever else you may find out about this change, you can know one fact for certain: the total electric charge on the gas will continue to be 50+.

Conservation of momentum. Two of the most useful conservation laws apply to the property known as momentum. Linear momentum is defined as the product of an object's mass and its velocity. (Velocity is the rate at which the position of an object changes with time, including both the speed and the direction.) A 200-pound football player moving with a speed of 10 miles per hour has a linear momentum of 200 pounds × 10 miles per hour, or 2,000 pound-miles per hour. In comparison, a 100-pound sprinter running at a speed of 20 miles per hour has exactly the same liner momentum: 100 pounds × 20 miles per hour, or 2,000 pound-miles per hour.

Linear momentum is consumed in any change. For example, imagine a rocket ship about to be fired into space (Figure 1a). If the rocket ship is initially at rest, its speed is 0, so its momentum must be 0. No matter what its mass is, the linear momentum of the rocket is mass × 0 miles per hour = 0. The important fact that the conservation of linear momentum tells us is that, whatever else happens to the rocket ship, its final momentum will also be 0.

What happens when the rocket is fired, then, as in Figure 1b? Hot gases escape from the rear of the rocket ship. The momentum of those gases is equal to their total mass (call that m_g) times their velocity (v_g), or $m_g v_g$. We'll give this number a negative sign ($-m_g v_g$) to indicate that the gases are escaping backward, or to the left.

The law of conservation of linear momentum says, then, that the rocket has to move in the opposite direction, to the right or the + direction, with a momentum of $m_g v_g$. That must be true because then $-m_g v_g$ (from the gases) plus $+m_g v_g$ (from the rocket) = 0. If you know the mass of the rocket, you can find the speed with which it will travel to the right.

Conservation laws

a. Before

Momentum of fuel ⟵ ⟶ Momentum of rocket

b. After

Figure 1. The rocket moves in the opposite direction of the escaping gases. *(Reproduced by permission of The Gale Group.)*

A second kind of momentum is angular momentum. The most familiar example of angular momentum is probably a figure skater spinning on the ice. The skater's angular momentum depends on three properties: her mass (or weight), the speed with which she is spinning, and the radius of her body.

At the beginning of a spin, the skater's arms may be extended outward, producing a large radius (the distance from the center of her body to the outermost part of his body). As she spins, she may pull her arms inward, bringing them to her side. What happens to the skater's angular momentum during the spin?

We can neglect the skater's mass, since she won't gain or lose any weight during the spin. The only factors to consider are the speed of her spin and her body radius. The law of conservation of angular momentum says that the product of these two quantities at the beginning of the spin ($v_1 r_1$) must be the same as the product of the two quantities at the end of the spin ($m_2 r_2$). So $m_1 r_1 = m_2 r_2$ must be true. But if the skater makes the radius of her body smaller, this equality can be true only if her velocity increases. This fact explains what you actually see on the ice. As a spinning skater pulls her arms in (and the body radius gets smaller), her spinning speed increases (and her velocity gets larger).

Conservation of parity. Conservation laws are now widely regarded as some of the most fundamental laws in all of nature. It was a great shock, therefore, when two American physicists, Val Lodgson Fitch (1923–) and James Watson Cronin (1931–), discovered in the mid-1960s that certain subatomic particles known as K-mesons appear to violate a conservation law. That law is known as the conservation of parity, which defines the basic symmetry of nature: that an object and its mirror image will behave the same way. Scientists have not yet fully explained this unexpected experimental result.

Constellation

A constellation is a group of stars that form a long-recognized pattern in the sky, as viewed from Earth. The stars that make up a constellation do not represent any meaningful order in the universe. Those stars may be at very different distances from Earth and from one another. Constellations seen from Earth would be shaped much differently and would be unrecognizable if viewed from another part of our galaxy.

> ## Words to Know
>
> **Asterism:** Familiar star pattern that is not a constellation.
>
> **Celestial sphere:** The sky or imaginary sphere that surrounds Earth and provides a visual surface on which astronomers plot celestial objects and chart their apparent movement due to Earth's rotation.
>
> **Ecliptic:** The apparent path of the Sun, the Moon, and the major planets among the stars in one year, as viewed from Earth.

The naming of constellations dates back to ancient civilizations. Although some constellations may resemble the animals, objects, or people for which they were named, others were merely named in honor of those figures. Many of the constellations were originally given Greek names and are related to ancient mythology. These names were later replaced by their Latin equivalents, names by which they are still known today.

Stargazing, however, was not limited to the ancient Greeks and Romans. Many cultures looked to celestial bodies to understand the creation

The constellation Orion, the Great Hunter. The three closely placed stars just left of center in this photo mark Orion's belt. *(Reproduced courtesy of National Aeronautics and Space Administration.)*

Continental margin

and structure of the universe and their place in it. Their naming of the different stars reflects their views or mythology. For example, the constellations the Romans called Ursa Major and Cassiopeia (pronounced kas-ee-o-PEE-a) were called Whirling Man and Whirling Woman by the Navajo.

Some familiar star groups known by common names are not constellations at all. These groups are called asterisms. Two examples are the Big Dipper and the Little Dipper. The Big Dipper, also known as the Plough, is part of the constellation Ursa Major (the Great Bear). The Little Dipper is part of the constellation Ursa Minor.

Eighty-eight constellations encompass the present-day celestial sphere (the sky or imaginary sphere that surrounds Earth). Each of these constellations is associated with a definite region in the celestial sphere. The yearly path of the Sun, the Moon, and the major planets among the stars, as viewed from Earth, is called the ecliptic. Twelve constellations are located on or near the ecliptic. These constellations—Aries, Taurus, Gemini, Cancer, Leo, Virgo, Libra, Scorpio, Sagittarius, Capricornus, Aquarius, and Pisces—are known as the constellations of the zodiac. The remaining constellations can be viewed in the celestial sphere during the year from either the Northern Hemisphere (28 constellations) or the Southern Hemisphere (48 constellations).

The daily rotation of Earth on it axis causes the constellations to appear to move westward across the sky each night. The yearly revolution of Earth around the Sun, which brings about the seasons, causes different constellations to come into view during the seasons.

[*See also* **Star**]

Continental margin

The continental margin is that underwater plain connected to continents, separating them from the deep ocean floor. The continental margin is usually divided into three major sections: the continental shelf, the continental slope, and the continental rise.

Continental shelf

Continental shelves are the underwater, gradually sloping ledges of continents. They tend to be quite flat, with an average seaward slope of less than 10 feet per mile (about 3 meters per kilometer). They vary in width from almost zero to more than 930 miles (1,500 kilometers), with a worldwide average of about 50 miles (80 kilometers). The widest shelves

> ## Words to Know
>
> **Continental rise:** A region at the base of the continental slope in which eroded sediments are deposited.
>
> **Continental shelf:** A gently sloping, submerged ledge of a continent.
>
> **Continental shelf break:** The outer edge of the continental shelf, at which the ocean floor drops off quite sharply in the continental slope.
>
> **Continental slope:** A steeply sloping stretch of the ocean that reaches from the outer edge of the continental shelf to the continental rise and deep ocean bottom.
>
> **Submarine canyon:** A steep V-shaped feature cut out of the continental slope by underwater rivers known as turbidity currents.
>
> **Turbidity current:** An underwater movement of water, mud, and other sediments.

are in the Arctic Ocean off the northern coasts of Siberia and North America. Narrow shelves are found off the western coasts of North and South America. The average depth at which the continental shelf begins to fall off toward the ocean floor (the beginning of the continental slope) is about 430 feet (130 meters).

Changes in sea level during Earth's history have alternatingly exposed and then covered portions of the continental shelf. During lowered sea level, land plants and animals, including humans and their ancestors, lived on the shelf. Today, their remains are often found there. For example, 12,000-year-old bones of mastodons, extinct relatives of the elephant, have been recovered off the coast of the northeastern United States.

Vast deposits of muds, sands, and gravels compose the continental shelf. Most commercial fishing takes place in the rich waters above the shelf. Many nations around the world claim ownership of the extensive oil, natural gas, mineral, and other natural resource deposits beneath the continental shelf adjacent to their land areas. Many nations also dump much of their waste in the ocean over the continental shelves.

Continental slope

At the seaward edge of the continental shelf is an immense drop-off. The steep edge where this occurs is known as the continental slope.

Contraception

The break point between the shelf and slope is sometimes known as the continental shelf break. The continental slopes are the most dramatic cliffs on the face of Earth. They may drop from a depth of 650 feet (200 meters) to more than 10,000 feet (3,000 meters) over a distance of 60 miles (100 kilometers). In the area of ocean trenches, the drop-off may be even more severe, from 650 feet (200 meters) to more than 33,000 feet (10,000 meters). In general, the steepest slopes tend to be found in the Pacific Ocean, and the least steep slopes in the Atlantic and Indian Oceans.

Submarine canyons. The most distinctive features of the continental slopes are submarine canyons. These are V-shaped features, often with tributaries, similar to canyons found on dry land. The deepest of the submarine canyons easily rival the size of the Grand Canyon of the Colorado River. Submarine canyons are created by the eroding flow of underwater rivers that travel across the continental slopes (and sometimes the continental shelf) carrying with them sediments that originated on the continents. These rivers are known as turbidity currents.

Continental rise

Sediments eroded off continental land, after being carried across the shelf and down the continental slope, are finally deposited at the base of the slope in a region of the ocean known as the continental rise. The deep ocean floor begins at the seaward edge of the rise. By some estimates, half of all the sediments laid down on the face of the planet are found in the gently sloping, smooth-surfaced continental rises.

[*See also* **Ocean**]

Contraception

Contraception, also called birth control, is the deliberate effort to halt conception a child (to keep a woman from becoming pregnant). Attempts to prevent pregnancy date back to ancient times and cultures. Some form of contraception is used by more than half the women in the United States. Although widespread, contraception remains controversial, with some religious and political groups opposed to distribution of contraceptives.

Ancient methods in use today

Some early methods of contraception involved techniques still used today. Gum arabic—a substance with which Egyptians coated tampons

Words to Know

Fallopian tube: One of a pair of structures in the female reproductive system that carries eggs from the ovaries to the uterus.

Fertilization: The union of an egg and sperm to form a new individual.

Hormone: A chemical messenger or substance produced by the body that has an effect on organs in other parts of the body.

Ovary: One of a pair of female reproductive organs that produces eggs and female sex hormones.

Ovulation: The release of an egg, or ovum, from an ovary.

Ovum: A mature female sex cell produced in the ovaries.

Sperm: A mature male sex cell secreted in semen during male ejaculation.

Uterus: The female organ in which the fetus develops before birth.

to kill sperm—is used to make spermicides contained in modern contraceptive jellies and foams. The ancient practice of prolonged nursing of infants to prevent conception of future children remains in current use, although it is by no means 100 percent effective. The modern diaphragm has its origin in a device made from bamboo that Asian women used as a barrier to the cervix (the opening to the uterus, or womb). The Chinese promoted "coitus interruptus," the withdrawal of the man's penis from the woman's vagina before ejaculation. Probably the most common contraceptive method in the world, this practice has resulted in numerous accidental pregnancies. The rhythm method (in which intercourse is avoided on the days of the month when a woman is most likely to become pregnant) was and remains the only form of birth control approved by the Roman Catholic Church.

Evolution of the condom

The practice of using condoms to prevent pregnancy and sexually transmitted diseases began in the sixteenth century, when cloth condoms were promoted to protect against syphilis. By the eighteenth century, condoms were made of animal membrane, making them waterproof and more effective as birth control devices. Latex (rubber) condoms were first pro-

Contraception

duced during the Industrial Revolution (about 1750 to about 1850). The emergence of acquired immunodeficiency syndrome (AIDS) in the 1980s again resulted in the widespread promotion of condom use as an effective barrier to disease.

Modern methods of contraception

Contraceptive devices that were developed in the late nineteenth century and are still used today include the diaphragm, a rubber cap that fits over the cervix and prevents the passage of sperm into the uterus; the contraceptive sponge, also a device used to cover the cervix before sexual intercourse; and foams and jellies containing spermicides that are inserted into the vagina before intercourse.

Advances in medical knowledge led to the development in the 1960s of the IUD (or intrauterine device), which is placed in the uterus to prevent or interrupt the process of conception. Birth control pills, approved for use in 1960 and the most popular contraceptive in the United States, contain hormones that are released into a woman's system on a regular basis (some are taken 21 days per month, others are taken every day) to prevent pregnancy. Different pills act in different ways: some inhibit ovulation (the release of an egg from the ovary), some prevent implantation

A variety of contraceptive methods. *(Reproduced by permission of The Stock Market.)*

of a fertilized egg (thereby denying cells the nourishment they need to develop into an embryo), and some thicken the secretions throughout the woman's reproductive system so that her partner's sperm has less of a chance to meet her egg.

Other recent developments include a matchsticklike device that is implanted under the skin of a woman's upper arm; it lasts about five years, releasing a contraceptive into the bloodstream that inhibits ovulation. An injectable form of contraceptive provides protection from pregnancy for three months at a time, but the most common reported side effects—including significant weight gain and mood swings—make this an unattractive choice for many women. In addition, a condom that can be inserted into the vagina of females became available in the mid-1990s, but its effectiveness is still being debated.

In 2000, in a landmark decision that received both widespread praise and protest, the U.S. Food and Drug Administration (FDA) approved the marketing of an abortion-inducing pill. This was the first alternative to surgical abortion approved in the United States. The prescription drug, called mifepristone or RU-486, was first developed in France in 1980. As of the end of 2000, 16 countries around the world had approved its use.

An abortion using mifepristone takes place in three steps. First, in a doctor's office, a woman is given a pregnancy test. If she is pregnant and it has been no longer than seven weeks since her last menstrual period, she is given three pills of mifepristone. The drug blocks the hormone progesterone, which is required to maintain a pregnancy. The woman then returns to the doctor's office within two days to take two tablets of a second drug, misoprostol. This second drug is a hormonelike substance that causes a woman's uterus to contract, expelling the fetal tissue, usually within six hours of taking the drug. Fourteen days later, she returns to her doctor's office and is checked to make sure she is no longer pregnant and no fetal tissue remains in her uterus. About 5 percent of the time, the abortion is incomplete and a woman will have to have a surgical abortion. Mifepristone fails completely in about 1 percent of the women who take it. The side effects of this abortion procedure are similar to a spontaneous miscarriage: uterine cramping, bleeding, nausea, and fatigue.

Sterilization

Sterilization, the surgical alteration of a male or female to prevent them from bearing children, is the most common form of birth control for women in the United States. In men, the operation is called a vasectomy. It is a simple out-patient procedure that involves snipping the vessel through which sperm passes so that semen—the off-white secretion ejected from the penis at the time of sexual climax—no longer contains sperm.

In women, sterilization involves a procedure called tubal ligation, in which the fallopian tubes that carry eggs from the ovaries to the uterus are tied or clipped. An egg that is released by an ovary during ovulation does not reach the uterus, thus preventing fertilization.

Challenges of contraception

Developing a foolproof method of birth control that has little or no side effects, is simple to use, and is agreeable to both men and women is a challenge. Sterilization is such a method, but only if the person undergoing the operation no longer wants to bear children.

Unwanted pregnancies can be measured by the rate of abortion (the ending of a pregnancy). Although many women who undergo abortions do not practice birth control, some pregnancies are the result of contraceptive failure. Abortion rates typically are highest in countries where contraceptives are not readily available. Some experts believe that easier access to contraceptive services would result in lower rates of accidental pregnancy and abortion.

[*See also* **Fertilization; Reproduction**]

Coral

Corals are a group of small, tropical marine animals that attach themselves to the seabed and form extensive reefs, commonly in shallow, warm-water seas. These reefs are made up of the calcium-carbonate (limestone) skeletons of dead coral animals. Coral reefs form the basis of complex marine food webs that are richer in species than any other ecosystem (community of plants and animals).

Biology of corals

A coral, or polyp, lives inside a cup-shaped skeleton that it secretes around itself. Resembling a sea anemone, a coral is a jelly-like sac attached at one end in its skeleton. The open end, the mouth, is fringed with stinging tentacles. A coral feeds by sweeping the water with its tentacles and stunning microscopic prey, which it then draws inside itself. Individual corals that gather together in large colonies are usually under one-eighth inch (3 millimeters) long. Living corals are often beautifully colored.

Corals reproduce two ways. Fertilized eggs released by the corals hatch to form larvae. After settling on a suitable surface, the larvae secretes its own limestone cup and grows into a mature coral, thus beginning a new colony. Corals also reproduce by budding, or forming new corals attached to themselves by thin sheets of tissue and skeletal material. In this way, corals grow into large, treelike structures.

Formation of coral reefs

Coral reefs are formed mainly by the hard skeletons of corals and the limestone deposits of coralline algae and other marine animals. Reefs grow upward as generations of corals produce limestone skeletons, die, and become the base for a new generation. Coral reefs lie in a zone of water 30°N to 30°S of the equator. Reef-forming coral animals flourish only in water under 100 feet (30 meters) deep and warmer than 72°F (22°C).

Coral reefs are classified into three main types. Fringing reefs grow close to the shore of a landmass, extending out like a submerged platform. Barrier reefs also follow a coastline, but are separated from it by wide expanses of water. Atolls are ring-shaped reefs surrounding lagoons.

The Great Barrier Reef of northeast Australia is the largest structure on Earth created by a living thing. It is 10 to 90 miles (16 to 145 kilometers)

Brightly colored orange cup coral found in western Mexico. *(Reproduced by permission of JLM Visuals.)*

Coral

wide and about 1,250 miles (2,010 kilometers) long, and is separated from the shore by a lagoon 10 to 150 miles (16 to 240 kilometers) wide.

Ecology of coral reefs and the damage caused by humans

With it numerous crevices and crannies, a coral reef is a home and feeding ground for countless numbers of fascinating marine life-forms. No ecosystem on Earth plays host to the diversity of inhabitants as found in and around a coral reef. Except for mammals and insects, almost every major group of animals is represented. More than 200 coral species alone are found in the Great Barrier Reef.

Coral reefs also benefit humans by protecting shorelines from the full onslaught of storm-driven waves. Humans, however, are responsible for causing severe damage to coral reefs. Reefs are often destroyed by collectors, who use coral to create jewelry, and fisherman, who use poison or dynamite to catch fish. Because corals need sunlight and sediment-free water to survive, water pollution poses a grave danger. Oil spills, the dumping of sewage wastes, and the runoff of soil and agricultural chemicals such as pesticides all threaten the delicately balanced ecosystem of coral reefs.

The extent of the damage done to the world's coral reefs was made clear by a report issued at the end of the year 2000. The Global Coral Reef

Kayangell atoll in Belau in the western Pacific Ocean. *(Reproduced by permission of Photo Researchers, Inc.)*

Monitoring Network, an international environmental monitoring organization, issued the report with data gathered from scientists around the globe. According to the report, the world has lost 27 percent of its coral reefs. Some of those reefs can never be recovered, while some could possibly come back. Most of the damaged reefs were found in the Persian Gulf, the Indian Ocean, the waters around Southeast and East Asia, and the Caribbean and adjacent Atlantic. The report pointed out that global warming was the biggest threat facing coral reefs, followed by water pollution, sediment from coastal development, and destructive fishing techniques (such as using dynamite and cyanide). If nothing is done to stop the destruction caused by humans, 60 percent of the world's coral reefs will disappear by 2030.

Correlation

As used in mathematics, correlation is a measure of how closely two variables change in relationship to each other. For example, consider the variables height and age for boys and girls. In general, one could predict that the older a child is, the taller he or she will be. A baby might be 12 inches long; an 8-year-old, 36 inches; and a 15-year old, 60 inches. This relationship is called a positive correlation because both variables change in the same direction: as age increases, so does height.

A negative correlation is one in which variables change in the opposite direction. An example of a negative correlation might be grades in school and absence from class. The more often a person is absent from class, the poorer his or her grades are likely to be.

The two variables compared to each other in a correlation are called the independent variable and the dependent variable. As the names suggest, an independent variable is one whose change tends to be beyond human control. Time is often used as an independent variable because it goes on whether we like it or not. In the simplest sense, time always increases, it never decreases.

A dependent variable is one that changes as the result of changes in the independent variable. In a study of plant growth, plant height might be a dependent variable. The amount by which a plant grows depends on the amount of time that has passed.

Correlation coefficient

Statisticians have invented mathematical devices for measuring the amount by which two variables are correlated with each other. The

Correlation

> **Words to Know**
>
> **Correlation coefficient:** A numerical index of a relationship between two variables.
>
> **Negative correlation:** Changes in one variable are reflected by changes in the second variable in the opposite direction.
>
> **Positive correlation:** Changes in one variable are reflected by similar changes in the second variable.

correlation coefficient, for example, ranges in value from -1 to $+1$. A correlation coefficient of $+1$ means that two variables are perfectly correlated with each other. Each distinct increase or decrease in the independent variable is accompanied by an exactly similar increase or decrease in the dependent variable. A correlation coefficient of $+0.75$ means that a change in the independent variable will be accompanied by a comparable increase in the dependent variable a majority of the time. A correlation coefficient of 0 means that changes in the independent and dependent variable appear to be random and completely unrelated to each other. And a negative correlation coefficient (such as -0.69) means that two variables respond in opposite directions. When one increases, the other decreases, and vice versa.

Understanding the meaning of correlation

It is easy to misinterpret correlational measures. They tell us nothing at all about cause and effect. For example, suppose that you measured the annual income of people from age 5 to age 25. You would probably find the two variables—income and age—to be positively correlated. The older people become, the more money they are likely to earn.

The wrong way to interpret that correlation is to say that growing older *causes* people to earn more money. Of course, that isn't true. The correlation can be explained in other ways. Obviously, a 5-year-old child can't earn money the way an 18-year-old or a 25-year-old can. Measures of correlation, such as the correlation coefficient, simply tell whether two variables change in the same way or not without providing any information as to the *reason* for that relationship.

Of course, scientists often design an experiment so that a measure of correlation *will* have some meaning. A nutrition experiment might be designed to test the effect of feeding rats a certain kind of food. The experimenter may arrange conditions so that only one factor—the amount of that kind of food—changes in the experiment. Every other condition is left the same throughout the experiment. In such a case, the amount of food is the independent variable and changes in the rat (such as weight changes) are considered the dependent variable. Any correlation between these two variables might then suggest (but would not prove) that the food being tested caused weight changes in the rat.

Cosmic ray

Cosmic rays are invisible, highly energetic particles of matter reaching Earth from all directions in space. Physicists divide cosmic rays into two categories: primary and secondary. Primary cosmic rays originate far outside Earth's atmosphere. Secondary cosmic rays are particles produced within Earth's atmosphere as a result of collisions between primary cosmic rays and molecules in the atmosphere.

Discovery of cosmic rays

The existence of cosmic radiation (energy in the form of waves or particles) was first discovered in 1912 by Austrian-American physicist Victor Hess during a hot-air balloon flight. Hess was trying to measure the background radiation that seemed to come from everywhere on the ground. The higher he went in the balloon, however, the more radiation he found. Hess concluded that there was radiation coming into our atmosphere from outer space.

Although American physicist Robert A. Millikan named these energy particles "cosmic rays" in 1925, he did not known what they were made of. In the decades since, physicists have learned much about cosmic rays, but their origin remains a mystery.

The nature of cosmic rays

An atom of a particular element consists of a nucleus surrounded by a cloud of electrons, which are negatively charged particles. The nucleus is made up of protons, which have a positive charge, and neutrons, which have no charge. These particles can be broken down further into smaller

Words to Know

Electron: A negatively charged particle, ordinarily occurring as part of an atom.

Electron volt (eV): The unit used to measure the energy of cosmic rays.

Neutron: Particle in the nucleus of an atom that possesses no charge.

Nucleus: The central mass of an atom, composed of neutrons and protons.

Photon: Smallest individual unit of electromagnetic radiation.

Primary cosmic ray: Cosmic ray originating outside Earth's atmosphere.

Proton: Positively charged particle composing part of the nucleus of an atom. Primary cosmic rays are mostly made up of single protons.

Radiation: Energy in the form of waves or particles.

Secondary cosmic ray: Cosmic ray originating within Earth's atmosphere as a result of a collision between a primary cosmic ray and some other particle or molecule.

Shower: Also air shower or cascade shower; a chain reaction of collisions between cosmic rays and other particles, producing more cosmic rays.

Subatomic particle: Basic unit of matter and energy smaller than an atom.

---

elements, which are called subatomic particles. Cosmic rays consist of nuclei and various subatomic particles. Most cosmic rays are protons that are the nuclei of hydrogen atoms. The nuclei of helium atoms, made up of a proton and a neutron, are the next common elements in cosmic rays. Together, hydrogen and helium nuclei make up about 99 percent of the primary cosmic radiation.

Primary cosmic rays enter Earth's atmosphere at a rate of 90 percent the speed of light, or about 167,654 miles (269,755 kilometers) per second. They then collide with gas molecules present in the atmosphere. These collisions result in the production of secondary cosmic rays of photons, neutrons, electrons, and other subatomic particles. These particles in turn collide with other particles, producing still more secondary radiation. When this cascade of collisions and particle production is quite extensive, it is known as a shower, air shower, or cascade shower. Secondary cosmic rays shower down to Earth's surface and even penetrate it.

Primary cosmic rays typically have energies that measure in the billions of electron volts (abbreviated eV). Energy is lost in collisions with other particles, so secondary cosmic rays are typically less energetic than primary ones. As the energies of the particles decrease, so do the showers of particles through the atmosphere.

The origin of cosmic rays

The ultimate origin of cosmic radiation is still not completely understood. Some of the radiation is believed to have been produced in the big bang at the origin of the universe. Low-energy cosmic rays are produced by the Sun, particularly during solar disturbances such as solar flares. Exploding stars, called supernovas, are also believed to be a source of cosmic rays.

[*See also* **Big bang theory; Particle detectors**]

One of the first cloud chamber photographs showing the track of a cosmic ray. It was taken by Dmitry Skobeltzyn in his laboratory in Leningrad in the Soviet Union in 1927. *(Reproduced by permission of Photo Researchers, Inc.)*

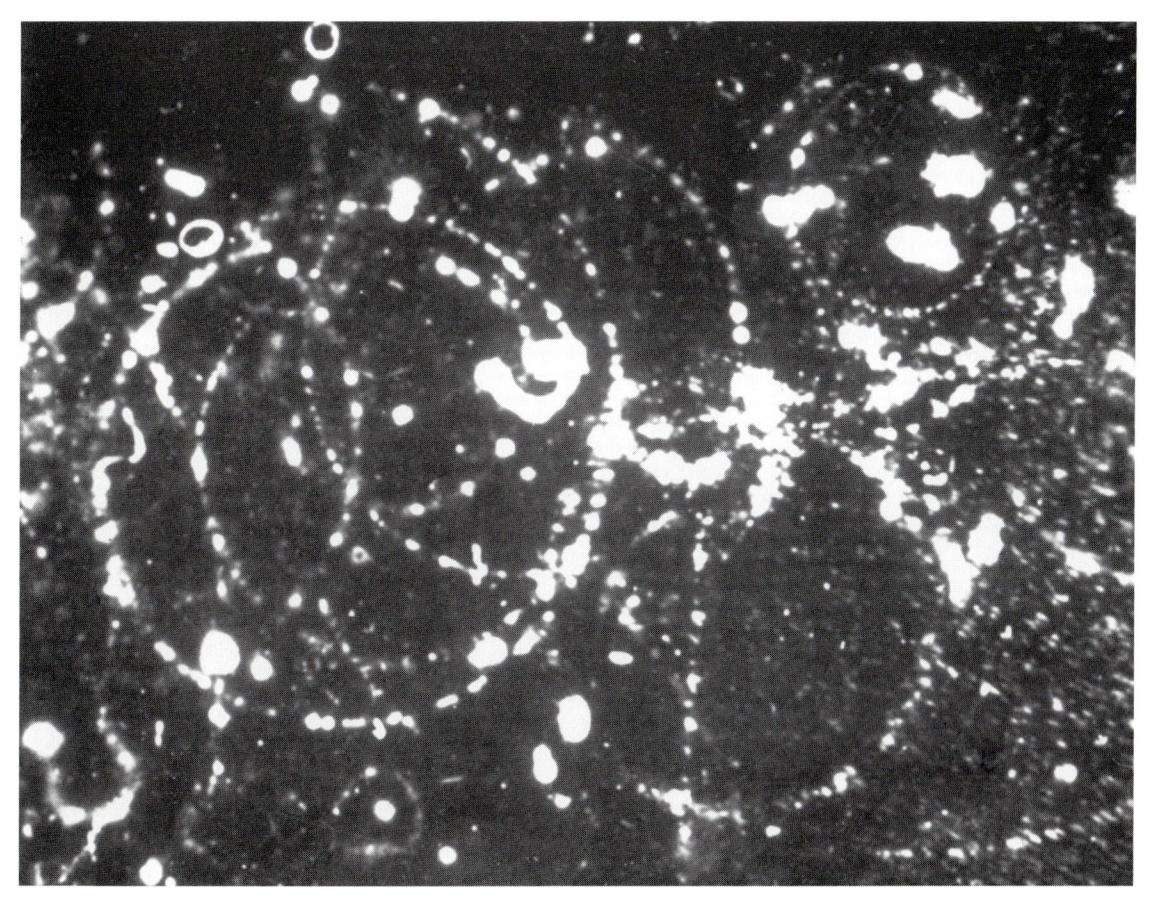

Cosmology

Cosmology is the study of the origin, evolution, and structure of the universe. This science grew out of mythology, religion, and simple observations and is now grounded in mathematical theories, technological advances, and space exploration.

Ancient astronomers in Babylon, China, Greece, Italy, India, and Egypt made observations without the assistance of sophisticated instruments. One of their first quests was to determine Earth's place in the universe. In A.D. 100, Alexandrian astronomer Ptolemy suggested that everything in the solar system revolved around Earth. His theory, known as the Ptolemaic system (or geocentric theory), was readily accepted (especially by the Christian Church) and remained largely unchallenged for 1,300 years.

A Sun-centered solar system

In the early 1500s, Polish astronomer Nicolaus Copernicus (1473–1543) rose to challenge the Ptolemaic system. Copernicus countered that the Sun was at the center of the solar system with Earth and the other planets in orbit around it. This sun-centered theory, called the Copernican system (or heliocentric theory), was soon supported with proof by Danish astronomer Tycho Brahe (1546–1601) and German astronomer Johannes Kepler (1571–1630). This proof consisted of careful calculations of the positions of the planets. In the early 1600s, Kepler developed the laws of planetary motion, showing that the planets follow an ellipse, or an oval-shaped path, around the Sun. He also pointed out that the universe was bigger than previously thought, although he still had no idea of its truly massive size.

The first astronomer to use a telescope to study the skies was Italian Galileo Galilei (1564–1642). His observations, beginning in 1609, supported the Copernican system. In the late 1600s, English physicist Isaac Newton (1642–1727) introduced the theories of gravity and mass, explaining how they are both responsible for the planets' motion around the Sun.

Over the next few centuries, astronomers and scientists continued to make additions to people's knowledge of the universe. These included the discoveries of nebulae (interstellar clouds) and asteroids (small, rocky chunks of matter) and the development of spectroscopy (the process of breaking down light into its component parts).

Words to Know

Asteroid: Relatively small, rocky chunk of matter that orbits the Sun.

Copernican system: Theory proposing that the Sun is at the center of the solar system and all planets, including Earth, revolve around it.

Gravity: Force of attraction between objects, the strength of which depends on the mass of each object and the distance between them.

Light-year: Distance light travels in one solar year, roughly 5.9 trillion miles (9.5 trillion kilometers).

Mass: Measure of the total amount of matter in an object.

Nebula: Cloud of interstellar gas and dust.

Ptolemaic system: Theory proposing that Earth is at the center of the solar system and the Sun, the Moon, and all the planets revolve around it.

Radiation: Energy in the form of waves or particles.

Spectroscopy: Process of separating the light of an object (generally, a star) into its component colors so that the various elements present within that object can be identified.

Speed of light: Speed at which light travels in a vacuum: 186,282 miles (299,728 kilometers) per second.

Modern cosmology

During the first two decades of the twentieth century, physicists and astronomers looked beyond our solar system and our Milky Way galaxy, forming ideas about the very nature of the universe. In 1916, German-born American physicist Albert Einstein (1879–1955) developed the general theory of relativity, which states that the speed of light is a constant and that the curvature of space and the passage of time are linked to gravity. A few years later, Dutch astronomer Willem de Sitter (1872–1934) used Einstein's theory to suggest that the universe began as a single point and has continued to expand.

In the 1920s, American astronomer Edwin Hubble (1889–1953) encountered observable proof that other galaxies exist in the universe besides our Milky Way. In 1929, he discovered that all matter in the

Cosmology

universe was moving away from all other matter, proving de Sitter's theory that the universe was expanding.

Creation of the universe

Astronomers have long been interested in the question of how the universe was created. The two most popular theories are the big bang theory and the steady-state theory. Belgian astrophysicist Georges-Henri Lemaître (1894–1966) proposed the big bang theory in 1927. He suggested that the universe came into being 10 to 15 billion years ago with a big explosion. Almost immediately, gravity came into being, followed by atoms, stars, and galaxies. Our solar system formed 4.5 billion years ago from a cloud of dust and gas.

In contrast, the steady-state theory claims that all matter in the universe has been created continuously, a little at a time at a constant rate, from the beginning of time. The theory, first elaborated in 1948 by Austrian-American astronomer Thomas Gold, also states that the universe is structurally the same all over and has been forever. In other words, the universe is infinite, unchanging, and will last forever.

Astronomers quickly abandoned the steady-state theory when microwave radiation (energy in the form of waves or particles) filling space throughout the universe was discovered in 1964. The existence of this radiation—called cosmic microwave background—had been predicted by supporters of the big bang theory. In April 1992, NASA (National Aeronautics and Space Administration) announced that its Cosmic Background Explorer (COBE) satellite had detected temperature fluctuations in the cosmic microwave background. These fluctuations indicated that gravitational disturbances existed in the early universe, which allowed matter to clump together to form large stellar bodies such as galaxies and planets. This evidence all but proves that a big bang is responsible for the expansion of the universe.

Continued discoveries

At the end of the twentieth century, astronomers continued to revise their notion of the size of the universe. They repeatedly found that it is larger than they thought. In 1991, astronomers making maps of the universe discovered great "sheets" of galaxies in clusters and super-clusters filling areas hundreds of millions of light-years in diameter. They are separated by huge empty spaces of darkness, up to 400 million light-years across. And in early 1996, the Hubble Space Telescope photographed at least 1,500 new galaxies in various stages of formation.

> ### Creationism
>
> Creationism is a theory about the origin of the universe and all life in it. Creationism holds that Earth is perhaps less than 10,000 years old, that its physical features (mountains, oceans, etc.) were created as a result of sudden calamities, and that all life on the planet was miraculously created as it exists today. It is based on the account of creation given in the Old Testament of the Bible.
>
> Because Creationism is not based on any presently held scientific principles, members of the scientific community dismiss it as a possible theory on how the universe was created. However, people who strongly believe in Creationism feel that it should be taught as a part of science education. The heated debate between the two sides continues.

In the late 1990s, while studying a certain group of supernovas, astronomers discovered that older objects in the group were receding at a speed similar to younger objects. In a "closed" universe, the expansion of the universe should slow down as it ages. Older supernovas should be receding more rapidly than younger ones. This is the theory that astronomers used to put forth: that at some future point the universe would stop expanding and then close back in on itself, an inverted big bang. However, with this recent finding, astronomers tend to believe that the universe is "open," meaning that the universe will continue its outward expansion for billions of years until everything simply burns out.

[*See also* **Big bang theory; Dark matter; Doppler effect; Galaxy; Redshift; Relativity, theory of**]

Cotton

Cotton is a fiber obtained from various species of woody plants and is the most important and widely used natural fiber in the world. The leading cotton-producing countries are China (the world's biggest producer), the United States, India, Pakistan, Brazil, and Egypt. The world production of cotton in the early 1990s was about 21 million tons (19 million metric tons) per year. The world's largest consumers of cotton are the United States and Europe.

Cotton

History

Cotton was one of the first cultivated plants, and it has been a part of human culture since prehistoric times. There is evidence that the cotton plant was cultivated in India as long as 5,000 years ago. Specimens of cotton cloth as old as 5,000 years have been found in Peru, and scientists have found 7,000-year-old specimens of the cotton plant in caves near Mexico City, Mexico.

Cotton plant

Cotton is primarily an agricultural crop, but it can also be found growing wild. There are more than 30 species of cotton plants, but only 4 are used to supply the world market for cotton. The cotton plant grows to a height of 3 to 6 feet (0.9 to 1.8 meters), depending on the species and the region where it is grown. The leaves are heart-shaped, lobed, and coarse veined, somewhat resembling a maple leaf. The plant has many branches with one main central stem. Overall, the plant is cone- or pyramid-shaped.

The seeds of the cotton plant are contained in capsules, or bolls. Each seed is surrounded by 10,000 to 20,000 soft fibers, white or creamy

Cotton plants in cultivation in North Carolina. *(Reproduced by permission of JLM Visuals.)*

in color. After the boll matures and bursts open, the fibers dry out and become tiny hollow tubes that twist up, making the fiber very strong.

Growing, harvesting, and processing

Cotton requires a long growing season (from 180 to 200 days), sunny and warm weather, plenty of water during the growth season, and dry weather for harvest. Cotton grows near the equator in tropical and semitropical climates. The cotton belt in the United States reaches from North Carolina down to northern Florida and west to California. Cotton plants are subject to numerous insect pests, including the destructive boll weevil.

For centuries, harvesting was done by hand. Cotton had to be picked several times in the season because bolls of cotton do not all ripen at the same time. The cotton gin, created by American inventor Eli Whitney (1765–1825) in 1793, mechanized the process of separating seeds from fibers, revolutionizing the cotton industry.

Before going to the gin, harvested cotton is dried and put through cleaning equipment that removes leaves, dirt, twigs, and other unwanted material. After cleaning, the long fibers are separated from the seeds with a cotton gin and then packed tightly into 500-pound (225-kilogram) bales. Cotton is classified according to its staple (length of fiber), grade (color), and character (smoothness). At a textile mill, cotton fibers are spun into yarn and then woven or knitted into cloth. At an oil mill, cottonseed oil is extracted from cotton seeds for use in cooking oil, shortening, soaps, and cosmetics.

Coulomb

A coulomb (abbreviation: C) is the standard unit of charge in the metric system. It was named after French physicist Charles A. Coulomb (1736–1806), who formulated the law of electrical force that now carries his name. (A physicist is one who studies the science of matter and energy.)

Coulomb's law concerns the force that exists between two charged particles. Suppose that two ping-pong balls are suspended in the air by threads at a distance of two inches from each other. Then suppose that both balls are given a positive electrical charge. Since both balls carry the same electrical charge, they will tend to repel—or push away from—each other. How large is this force of repulsion?

Coulomb

> **Words to Know**
>
> **Electrolytic cell:** Any cell in which an electrical current is used to bring about a chemical change.
>
> **Proportionality constant:** A number that is introduced into a proportionality expression in order to make it into an equality.
>
> **Quantitative:** Any type of measurement that involves a mathematical measurement.
>
> **Torsion:** A twisting force.

History

The period between 1760 and 1780 was one in which physicists were trying to answer that very question. They already had an important clue as to the answer. A century earlier, English physicist Isaac Newton (1642–1727) had discovered the law of gravity. Two objects attract each other, that law says, with a force that depends on the masses of the two bodies and the distance between them. The law is an inverse square law. That is, as the distance between two objects doubles (increases by 2), the force between them decreases by one-fourth ($1 \div 2^2$). As the distance between the objects triples (increases by 3), the force decreases by one-ninth ($1 \div 3^2$). Perhaps, physicists thought, a similar law might apply to electrical forces.

The first experiments in this field were conducted by Swiss mathematician Daniel Bernoulli (1700–1782) around 1760. Bernoulli's experiments were apparently among the earliest studies in the field of electricity that used careful measurements. Unfamiliar with such techniques, however, most scientists paid little attention to Bernoulli's results.

About a decade later, two early English chemists—Joseph Priestley (1733–1804) and Henry Cavendish (1731–1810)—carried out experiments similar to those of Bernoulli. Priestley and Cavendish concluded that electrical forces are indeed similar to gravitational forces. But they did not discover a concise mathematical formula like Newton's.

The problem of electrical forces was finally solved by Coulomb in 1785. The French physicist designed an ingenious apparatus for measuring the small force that exists between two charged bodies. The apparatus is known as a torsion balance.

A torsion balance consists of two parts. One part is a horizontal bar made of a material that does not conduct electricity. Suspended from each end of the bar by means of a thin fiber of metal or silk is a ping-pong-like ball. Each of the two balls is given an electrical charge. Finally, a third ball is placed next to one of the balls hanging from the torsion balance. In this arrangement, a force of repulsion develops between the two adjacent balls (balls that are side by side). As they push away from each other, they cause the metal or silk fiber to twist. The amount of twist that develops in the fiber can be measured and can be used to calculate the force existing between the bodies.

Coulomb's law

The results of this experiment allowed Coulomb to write a mathematical equation for electrical force. The equation is similar to that for gravitational forces. Suppose that the charges on two bodies are represented by the letters q_1 and q_2, and the distance between them by the letter r. Then the electrical force between the two is proportional to q_1 times q_2 ($q_1 \times q_2$). It is also inversely proportional to the distance, or $1/r^2$.

The term inverse means that as one variable increases, the other decreases. As the distance between two charged particles increases, the force decreases. Furthermore, the change occurs in a square relationship. That is, as with gravitational forces, when the distances doubles (increases by 2), the force decreases by one-fourth (by $\frac{1^2}{2}$). When the distance triples (increases by 3), the force decreases by one-ninth (by $\frac{1^2}{3}$), and so on.

Electrical and magnetic forces are closely related to each other, so it is hardly surprising that Coulomb also discovered a similar law for magnetic force a few years later. The law of magnetic force says that it, too, is an inverse square law.

Applications

Coulomb's law is one of the basic laws of physics (the science of matter and energy). Anyone who studies electricity uses this principle over and over again. But Coulomb's law is used in other fields of science as well. One way to think of an atom, for example, is as a collection of electrical charges. Protons each carry one unit of positive electricity, and electrons carry one unit of negative electricity. (Neutrons carry no electrical charge and are, therefore, of no interest from an electrical standpoint.)

Therefore, chemists (who study atoms) have to work with Coulomb's law. How great is the force of repulsion among protons in an atomic nucleus? How great is the force between the protons and electrons in an atom?

How great is the electrical force between two adjacent atoms? Chemical questions like these can all be answered by using Coulomb's law.

Another application of Coulomb's law is in the study of crystal structure. Crystals are made of charged particles called ions. Ions arrange themselves in any particular crystal (such as a crystal of sodium chloride, or table salt) so that electrical forces are balanced. By studying these forces, mineralogists can better understand the nature of specific crystal structures.

Electrolytic cells

The coulomb (as a unit) can be thought of in another way, as given by the following equation: 1 coulomb = 1 ampere × 1 second. The ampere (amp) is the metric unit used for the measurement of electrical current. (Electrical appliances in the home operate on a certain number of amps.) One amp is defined as the flow of electrical charge per second of time. Thus, by multiplying the number of amps by the number of seconds that elapse, the total electrical charge (number of coulombs) can be calculated.

This information is of significance in the field of electrochemistry because of a discovery made by British scientist Michael Faraday (1791–1867) around 1833. Faraday discovered that a given quantity of electrical charge passing through an electrolytic cell will cause a given amount of chemical change in that cell. For example, if one mole of electrons flows through a cell containing copper ions, one mole of copper will be deposited on the cathode or electrode of that cell. (A mole is a unit used to represent a certain number of particles, usually atoms or molecules.) The Faraday relationship is fundamental to the practical operation of many kinds of electrolytic cells.

[*See also* **Electric current**]

Crops

Crops are plants or animals or their products cultivated (grown, tended, and harvested) by humans as a source of food, materials, or energy. Humans are rather particular in their choice of crops. Though they select a wide range of useful species of plants and animals to raise, there are vast diversities of species available in particular places or regions.

Farming can involve the cultivation of plants and livestock on farms, fish and other aquatic animals in aquaculture, and trees in agroforestry plantations.

Words to Know

Agroforestry: Cultivation of crops of trees under managed conditions, usually in single-species plantations.

Aquaculture: Managed breeding of aquatic animals and plants for use as food.

Fallow: Cultivated land that is allowed to lie idle during the growing season so that it can recover some of its nutrients and organic matter.

Organic matter: Remains, residues, or waste products of any living organism.

Plants

Hundreds of species of plants are cultivated by humans under managed conditions. However, a remarkably small number of species contribute greatly to the global harvest of plant crops. Ranked in order of their annual production, the world's 15 most important food crops are:

An orchard of olive trees in the Aegean coast of Turkey. *(Reproduced by permission of JLM Visuals.)*

Crops

> ## Foot-and-mouth disease
>
> Diseases and severe weather can easily destroy the crops—animals or plants—cultivated on a farm. The effect can be economically devastating. One such agricultural blight occurred in England in early 2001. In February of that year, on a farm in Essex in eastern England, 27 pigs contracted foot-and-mouth disease (also called hoof-and-mouth disease).
>
> This virus affects animals with hooves, such as cattle, pigs, sheep, goats, and deer. Elephants, rats, and hedgehogs are also susceptible. Animals afflicted with the disease suffer from fever, followed by the eruption of blisters in the mouth, on the hooves or feet, on tender skin areas such as the udder in females, and in the nostrils. The blisters grow large, then break, exposing raw surfaces. Eating becomes difficult and painful. Because the soft tissues under the afflicted animal's hooves are inflamed, the animal becomes lame and may even shed its hooves. Pregnant female animals often abort and dairy cattle may give less milk. The disease is rarely fatal, although it can cause death in very young animals. Only very rarely is the disease transmitted to humans. When it has been, only mild symptoms have appeared.
>
> Foot-and-mouth disease is a highly contagious virus spread by direct or indirect contact. It can be carried by birds, on clothes, on the tires of vehicles, on dust, and in infected meat eaten by animals. It can travel many miles simply borne on the wind. Once an animal is infected, symptoms begin to show one to ten days later. Heat, sun-

sugar cane, wheat, rice, corn (maize), white potatoes, sugar beets, barley, sweet potatoes, cassava, soybeans, wine grapes, tomatoes, bananas, legumes (beans and peas), and oranges.

Most farm crops are managed as annual plants, meaning they are cultivated in a cycle of one year or less. This is true of all of the grains and legumes and most vegetables. Other species, however, are managed as perennial plants, which once established are capable of yielding crops on a continued basis. This is typically the manner in which tree-fruit crops such as oranges are managed and harvested.

Some crops are cultivated to produce important medicines. An example is rosy periwinkle, which produces several chemicals that are ex-

Crops

light, and disinfectants destroy the virus. Scientists have made progress in developing a vaccine against the disease, but the cost of vaccinating all susceptible animals (possibly up to $1 billion a year) is seen by some public officials as prohibitive. Consequently, to prevent a widespread outbreak that could cause massive production losses, infected animals must be destroyed by incineration and affected areas must be isolated.

The most serious outbreak in the United States occurred in 1914 when animals in 22 states and the District of Columbia were stricken. The last major incidence of foot-and-mouth disease in England occurred in 1967. Some 440,000 animals were slaughtered, costing the farming industry $200 million (equivalent to $2.3 billion today).

Within weeks of the 2001 outbreak in England, hundreds of thousands of animals were infected and had to be killed and incinerated. Many others who were not infected were killed as a precautionary measure to stop the spread. The virus soon showed up in other European countries: France, the Netherlands, and Ireland all reported confirmed cases. Farming was not the only industry hit hard by the outbreak. Tourism was also severely affected as footpaths and walking trails in the English countryside that meandered near farms were closed to the public in an effort to halt the spread of the virus. The tourism industry lost an estimated $150 million a week as a result. The total cost to all industries affected in England (farming, sports, tourism) was estimated to be over $13 billion.

tremely useful in the treatment of certain types of cancers. Other crops are cultivated to produce extremely profitable but illegal drugs for unlawful markets. Examples of these sorts of crops include marijuana, coca, and opium poppy.

Animals

Enormous numbers of domesticated animals are raised by humans for use as food and materials. In many cases, the animals are used to produce some product that can be harvested without killing the animals. For example, milk can be continuously collected from various species of mammals, especially cows. Similarly, chickens can produce eggs regularly.

Mad-cow disease

Mad-cow disease (known properly as bovine spongiform encephalopathy or BSE) is another disease that can destroy a population of farm animals. Unlike foot-and-mouth disease, mad-cow disease primarily attacks cattle, and it is fatal. Even more, mad-cow disease has spawned a human form of the disease called variant Creutzfeldt-Jacob disease (vCJD). At present, vCJD is incurable and fatal.

Scientists believe mad-cow disease is caused not by a bacterium or virus but by a self-replicating, abnormally folded protein called a prion. Scientists are still unsure of the complete nature of this agent. It resists freezing, drying, and heating at normal cooking temperatures, even those used for pasteurization and sterilization. Once inside an animal, it incubates for a period of four or five years. Once the disease appears, the animal is dead within weeks or months. The abnormal protein affects the brain and spinal cord of cattle, converting normal proteins into the abnormal form. The accumulation of abnormal proteins in the brain is marked by the formation of spongy holes. Cows afflicted with the disease grow ill-tempered and wobbly, lose weight, then suffer seizures, paralysis, blindness, and, finally, death.

In 1984, on a farm in South Downs, England, a number of cows died from a strange malady that veterinarians could not identify. Two years later, the scientific community specifically diagnosed the malady as mad-cow disease. In 1988, officials in England ordered the destruction of stricken cows and banned the use of cows, sheep, and other various hoofed animals in cattle feed. However, the banned

However, these plus many other domesticated animals are routinely slaughtered for their meat. The populations of some of these domesticated animals are enormously large. In addition to 6 billion people, the world today supports about 1.7 billion sheep and goats; 1.3 billion cows; 0.9 billion pigs; and 0.3 billion horses, camels, and water buffalo. In addition, there are about 10 to 11 billion domestic fowl, most of which are chickens.

Aquaculture

Aquaculture refers to the managed breeding of aquatic animals and plants for use as food. Increasingly, aquaculture or fish farming is seen cattle feed was exported to other countries for another eight years. Between 1988 and 1996, nations in Asia alone bought nearly 1 million tons (0.9 million metric tons).

Between November 1986 and December 2000, approximately 180,000 cases of mad-cow disease were confirmed in England. The epidemic peaked in the period 1992 to 1993 with almost 1,000 cases reported a week. Since that time, the disease has been reported in domestic cattle in Ireland, France, Portugal, and Switzerland, and in a number of countries that received cattle feed from England. No cases of mad-cow disease have been found in the United States. England has banned the recycling of farm animals and has stopped exporting meat-based cattle feed. It has also spent billions of dollars destroying and disposing of cows, both those that were infected and those that were merely old.

The first person to develop symptoms of what turned out to be vCJD became ill in January 1994. Scientists now believe this brain-wasting malady, with symptoms similar to mad-cow disease, is caused by eating the meat of animals who suffered from mad-cow disease. The incubation period for vCJD is thought to be between ten and sixteen years. From the time it was first identified until December 2000, 87 cases of vCJD were reported in England, 3 in France, and 1 in the Republic of Ireland. Because of its long incubation period, scientists are unsure how many more cases of vCJD will arise around the world in the future.

as an alternative to the exploitation of wild species of aquatic animals and plants.

Most freshwater aquaculture occurs in inland areas where there are many ponds and small lakes. Freshwater aquaculture is particularly important in Asia, where various species of fish, especially carp and tilapia, are raised in artificial ponds. In North America, the most common species of fish grown in small ponds are rainbow trout and catfish.

Aquaculture is also becoming increasingly important along sheltered ocean coastlines in many parts of the world. In the tropics, extensive areas of mangrove forest are being converted into shallow ponds for the cultivation of prawns. In North America and Western Europe, the cultivation of Atlantic salmon became an extensive industry in the late twentieth century, using pens floating in shallow, coastal bays and sometimes in the open ocean.

Agroforestry

In agroforestry, trees are raised as a source of lumber, pulpwood, or fuel. In many regions, this sort of intensive forestry is being developed as an alternative to the harvesting of natural forests.

In temperate zones, the most important trees grown are species of pine, spruce, larch, and poplar. Depending on the species and site conditions, these trees can be harvested after a growth period of only 10 to 60 years, compared with 60 to more than 100 years for natural, unmanaged forests in the same regions.

In the tropics, fast-growing, high-yield species of trees are grown for use locally as fuel, animal fodder, lumber, and pulpwood. Various tree species include pine, eucalyptus she-oak, and tree-legumes. Slower-growing tropical hardwoods, such as mahogany and teak, are grown for their high-quality lumber.

Climate and crops

Climate dominates agriculture, second only to irrigation. Crops are especially vulnerable to weather variations, such as late or early frosts, heavy rains, or drought. Because of their ability to grow within a climate range, rice, wheat, and corn have become the dominant crops globally.

These crops all need a wet season for germination and growth, followed by a dry season to allow spoilage-free storage. Rice was domesticated in the monsoonal lands of Southeast Asia, while wheat originated in the Fertile Crescent of the Middle East. Historically, wheat was planted in the fall and harvested in late spring, coinciding with the cycle of wet and dry seasons in the Mediterranean region. Corn needs the heavy summer rains provided by the Mexican highland climate.

Other crops are prevalent in areas with less suitable climates. These include barley in semiarid lands (those with light rainfall); oats and potatoes in cool, moist lands; rye in colder climates with short growing seasons; and dry rice on hillsides and drier lands.

Although food production is the main emphasis in farming, more and more industrial applications have evolved. Cloth fibers, such as cotton, have been a mainstay, but paper products and many chemicals now come from cultivated plants.

Crop rotation

When farmers grow two or more crops alternately on the same land, they are rotating crops. Farmers rotate crops to control erosion; promote the fertility of the soil; and control weeds, insects, and plant diseases.

Farmers have been following the practice of crop rotation since the time of the ancient Romans. With the advent of chemical fertilizers following World War II (1939–45), crop rotation fell out of favor somewhat. Farmers could provide their crops with nutrients without leaving some of their lands fallow (plowed and tilled but unseeded) each year. But some farmers eventually returned to the practice of rotating crops as a way to improve the structure and health of their soils.

Farmers rotate crops to assure that the soil is covered for as much of the year as possible to protect it from exposure to water, wind, and other elements that cause erosion. On sloping lands, rotating crops increases the organic content (remains or residue of living organisms) and overall stability of the soil, further decreasing the chance for erosion. Another method farmers use to increase the organic matter of soil is to plant rotation crops that help build the structure of the soil, such as alfalfa, sweet clover, or red clover. When plowed under the soil, these crops decompose quickly, giving the soil a high level of nutrients.

Adjacent fields of rice and wheat in California's Sacramento Valley. *(Reproduced by permission of JLM Visuals.)*

Crustaceans

To control insects and weeds, farmers can rotate crops that have different characteristics: rotate weed-suppressing crops with those that do not suppress weeds and rotate crops susceptible to specific insects with those that are not. If the same crop is not grown in the same field one year after another, the reproductive cycles of insects preying on a specific plant are interrupted. Farmers can also control insects by planting next to each other certain crops that do not attract the same insects.

[*See also* **Agriculture; Agrochemicals; Aquaculture; Forestry; Organic farming**]

Crustaceans

The crustaceans are a group of animals that belong to the class Crustacea in the phylum Arthropoda (organisms with segmented bodies, jointed legs or wings, and an external skeleton). The class includes a wide variety of familiar animals, such as barnacles, crabs, crayfish, copepods, shrimp, prawns, lobsters, water fleas, and wood lice. More than 30,000 species of crustacea have been identified, the majority of which live in water. Species that live in moist habitats on land, such as wood lice and pill bugs, are believed to have evolved from marine species.

Most crustaceans are free-living but some species are parasitic. Some even live on other crustaceans. Some species are free-swimming, while others crawl or burrow in soft sediments.

General structural characteristics

Despite such an extraordinary diversity of species, many crustaceans have a similar structure and way of life. The distinctive head usually bears five pairs of appendages (limblike attachments). Two pairs of these appendages are antennae that are used to detect food as well as to sense changes in humidity and temperature. Another pair of appendages are mandibles (jaws) that are used for grasping and tearing food. The final two pairs of appendages are maxillae, armlike projections used for feeding purposes.

The main part of the body is taken up with the thorax and abdomen. Both of these segments are covered with a tough outer skeleton, or exoskeleton. The exoskeleton is generally harder than it is in other arthropods because it contains limestone in addition to chitin (pronounced KITE-in), the usual skeletal material.

Attached to the trunk region are a number of other appendages which vary both in number and purpose in different species. In crabs, for ex-

Words to Know

Appendage: A limblike attachment extending from the main part of an animal's body.

Arthropoda: The largest single animal phylum, consisting of organisms with segmented bodies, jointed legs or wings, and exoskeletons.

Dioecious: A type of animal that exists as either male or female.

Exoskeleton: An external skeleton.

Free-living: An organism that is able to move about in its search for food and is not attached to some other organism as, for example, a parasite.

Hermaphrodite: An organism with both male and female sex cells.

Larva: An immature form of an organism capable of surviving on its own.

Molt: The process by which an organism sheds its outermost layer of feathers, fur, skin, or exoskeleton.

Parasite: An organism that lives in or on a host organism and that gets its nourishment from that host.

Seta: A thin, whiskerlike projection extending from the body of an organism.

ample, one pair of appendages may be modified for swimming, another for feeding, another for carrying eggs, and yet another for catching prey.

Life patterns

Crustaceans exhibit a wide range of feeding techniques. The simplest of these techniques are used by species such as the tiny shrimps and copepods that practice filter feeding. In filter feeding, an animal rhythmically waves many fine oarlike structures known as setae back and forth. This motion creates a mini water current towards the mouth. Plankton and other suspended materials are carried into special filters and then transferred to the mouth.

Larger species such as crabs and lobsters are active hunters of small fish and other organisms. Other species adopt a scavenging role, feeding on dead animals or plants and other waste materials.

Crustaceans

Smaller forms of crustaceans obtain the oxygen they need by gas exchange through their entire body surface. Most crustaceans, however, have special gills that serve as a means of obtaining oxygen. Simple excretory organs provide for the removal of body wastes such as ammonia and urea. Most crustaceans have a series of well-developed sensory organs that include not only eyes, but also a range of chemical and tactile (touch) receptors. All crustaceans are probably capable of detecting a light source, but in some of the more developed species, definite shapes and movements may also be detected.

Breeding strategies vary considerably among crustaceans. Most species are dioecious (pronounced die-EE-shus), that is, are either male or female. But some, such as the barnacles, are hermaphroditic. An hermaphroditic animal is one that possesses both male and female sex organs. Fertilization usually occurs sexually between two individuals. The fertilized eggs then mature either in a specialized brood chamber in some part of the female's body or attached directly to some external appendage

Horseshoe crabs come ashore for annual mating and nesting in Delaware Bay. *(Reproduced by permission of The Stock Market.)*

such as a claw. Most aquatic species hatch into a free-swimming larvae that progress through a series of body molts (where they shed their skin) until finally arriving at the adult size.

Cryobiology

Cryobiology is the study of the effects of very low temperatures on living things. Cryobiology can be used to preserve, store, or destroy living cells. At very low temperatures, cellular metabolism (all the chemical reactions that drive the activities of cells) essentially stops. Freezing technology is used for food preservation, blood storage at hospitals and blood banks, sperm and egg storage, preservation of some transplant tissues, and certain delicate surgeries. Cryopreservation, the freezing and eventual thawing of living material, is the most advanced use of this technology.

History

Ice has been used to slow the decay of food for centuries, but the widespread industrial use of freezing has occurred only in about the last 100 years with advances in refrigeration and cryotechnology. By the 1940s, red blood cells were being frozen to provide blood supplies as needed to the wounded during World War II (1939–45). Not long afterward, farmers began freezing the sperm of bulls in order to impregnate cows at distant locations.

The cryopreservation of single cells or small clumps of cells has been carried out successfully. However, preservation of whole live organs by freezing is more difficult and—as of the beginning of the twenty-first century—have largely failed.

Cellular cryopreservation

Cryopreservation involves keeping cells at extremely low temperatures until they are needed. Careful steps must be taken to insure that the frozen cells will survive and be in good condition when they are thawed. It is important that cells be healthy prior to freezing. Cryopreservation places stress on cells that can cause even some healthy ones to perish during the freezing and thawing process.

Medical and surgical applications

The most successful medical applications of cryopreservation are in blood storage and the field of fertility. Blood banks can freeze rare and

Cryogenics

> ### Words to Know
>
> **Cryopreservation:** The preservation of cells by keeping them at extremely low temperatures.
>
> **Cryosurgery:** Surgery that involves the freezing of tissue to be treated or operated on.

individual blood types for up to 10 years. Patients who have leukemia (a type of cancer) and must undergo radiation treatments can have their sperm and bone marrow cells, which are radiation sensitive, frozen and stored for later use. Some men undergoing a vasectomy (a sterilization procedure) store sperm so that they have the option of fathering a child at a later date. In addition, fertilized eggs have been successfully frozen and thawed for placement into the female's uterus, or womb.

Medical scientists can also use freezing technology during surgical procedures to improve a patient's prognosis (the outcome of the surgery). In transplant surgery, eye or skin tissue to be transplanted is sometimes frozen before use. Cryosurgery is the freezing of unwanted tissue, such as precancerous lesions or warts, in order to destroy it. A major advantage of cryosurgery is that it produces less scarring than surgery in which the tissue is removed by cutting. Cold temperatures are also sometimes used to cool patients before an operation. When a patient's brain and heart are cooled, they need less oxygen, thereby permitting longer surgery.

[See also **Cryogenics**]

Cryogenics

Cryogenics is the science of producing and studying low-temperature conditions. The word cryogenics comes from the Greek word *cryos,* meaning "cold," combined with a shortened form of the English verb "to generate." It has come to mean the generation of temperatures well below those of normal human experience. More specifically, a low-temperature environment is termed a cryogenic environment when the temperature range is below the point at which permanent gases begin to liquefy. Permanent gases are elements that normally exist in the gaseous state and

> ### ▼ Words to Know
>
> **Absolute zero:** The lowest temperature possible at which all molecular motion ceases. It is equal to −273°C (−459°F).
>
> **Kelvin temperature scale:** A temperature scale based on absolute zero with a unit, called the kelvin, having the same size as a Celsius degree.
>
> **Superconductivity:** The ability of a material to conduct electricity without resistance. An electrical current in a superconductive ring will flow indefinitely if a low temperature (about −260°C) is maintained.

were once believed impossible to liquefy. Among others, they include oxygen, nitrogen, hydrogen, and helium.

The origin of cryogenics as a scientific discipline coincided with the discovery by nineteenth-century scientists that the permanent gases can be liquefied at exceedingly low temperatures. Consequently, the term "cryogenic" applies to temperatures from approximately −100°C (−148°F) down to absolute zero (the coldest point a material could reach).

The temperature of any material—solid, liquid, or gas—is a measure of the energy it contains. That energy is due to various forms of motion among the atoms or molecules of which the material is made. A gas that consists of very rapidly moving molecules, for example, has a higher temperature than one with molecules that are moving more slowly.

In 1848, English physicist William Thomson (later known as Lord Kelvin; 1824–1907) pointed out the possibility of having a material in which particles had ceased all forms of motion. The absence of all forms of motion would result in a complete absence of heat and temperature. Thomson defined that condition as absolute zero.

Thomson's discovery became the basis of a temperature scale based on absolute zero as the lowest possible point. That scale has units the same size as the Celsius temperature scale but called kelvin units (abbreviation K). Absolute zero is represented as 0 K, where the term degree is omitted and is read as zero kelvin. The Celsius equivalent of 0 K is −273°C, and the Fahrenheit equivalent is −459°F. One can convert between Celsius and Kelvin scales by one of the following equations:

$$°C = K - 273 \text{ or } K = °C + 273$$

Cryogenics

Cryogenics, then, deals with producing and maintaining environments at temperatures below about 173 K.

One aspect of cryogenics involves the development of methods for producing and maintaining very low temperatures. Another aspect includes the study of the properties of materials at cryogenic temperatures. The mechanical and electrical properties of many materials change very dramatically when cooled to 100 K or lower. For example, rubber, most plastics, and some metals become exceedingly brittle. Also many metals and ceramics lose all resistance to the flow of electricity, a phenomenon called superconductivity. In addition, helium that is cooled to very nearly absolute zero (2.2 K) changes to a state known as superfluidity. In this state, helium can flow through exceedingly narrow passages with no friction.

History

Cryogenics developed in the nineteenth century as a result of efforts by scientists to liquefy the permanent gases. One of the most successful of these scientists was English physicist Michael Faraday (1791–1867). By 1845, Faraday had managed to liquefy most permanent gases then known to exist. His procedure consisted of cooling the gas by immersion in a bath of ether and dry ice and then pressurizing the gas until it liquefied.

Six gases, however, resisted every attempt at liquefaction and were known at the time as permanent gases. They were oxygen, hydrogen, nitrogen, carbon monoxide, methane, and nitric oxide. The noble gases—helium, neon, argon, krypton, and xenon—were yet to be discovered. Of the known permanent gases, oxygen and nitrogen (the primary constituents of air), received the most attention.

For many years investigators labored to liquefy air. Finally, in 1877, Louis Cailletet (1832–1913) in France and Raoul Pictet (1846–1929) in Switzerland succeeded in producing the first droplets of liquid air. Then, in 1883, the first measurable quantity of liquid oxygen was produced by S. F. von Wroblewski (1845–1888) at the University of Krakow. Oxygen was found to liquefy at 90 K, and nitrogen at 77 K.

Following the liquefaction of air, a race to liquefy hydrogen ensued. James Dewar (1842–1923), a Scottish chemist, succeeded in 1898. He found the boiling point of hydrogen to be a frosty 20 K. In the same year, Dewar succeeded in freezing hydrogen, thus reaching the lowest temperature achieved to that time, 14 K. Along the way, argon was discovered (1894) as an impurity in liquid nitrogen. Somewhat later, krypton and xenon were discovered (1898) during the fractional distillation of liquid argon. (Fractional distillation is accomplished by liquefying a mixture of gases, each of which has a different boiling point. When the mixture is evaporated, the gas with the highest boiling point evaporates first, followed by the gas with the second highest boiling point, and so on.)

Each of the newly discovered gases condensed at temperatures higher than the boiling point of hydrogen but lower than 173 K. The last element to be liquefied was helium gas. First discovered in 1868 in the spectrum of the Sun and later on Earth (1885), helium has the lowest boiling point of any known substance. In 1908, Dutch physicist Heike Kamerlingh Onnes (1853–1926) finally succeeded in liquefying helium at a temperature of 4.2 K.

Methods of producing cryogenic temperatures

Cryogenic conditions are produced by one of four basic techniques: heat conduction, evaporative cooling, cooling by rapid expansion (the Joule-Thomson effect), and adiabatic demagnetization. The first two are well known in terms of everyday experience. The third is less well known but is commonly used in ordinary refrigeration and air conditioning units, as well as in cryogenic applications. The fourth process is used primarily in cryogenic applications and provides a means of approaching absolute zero.

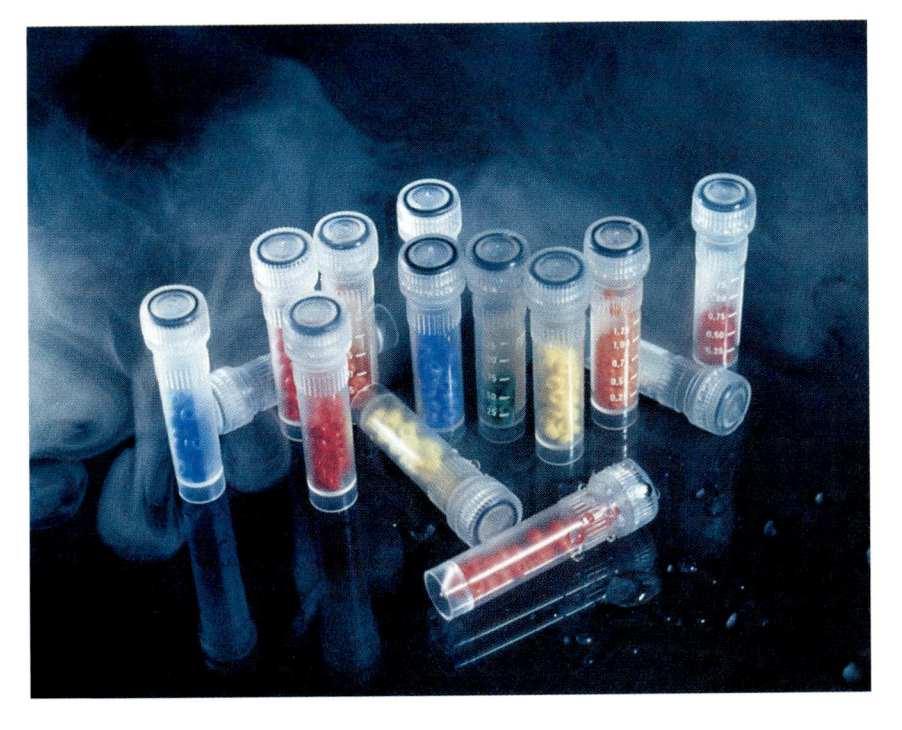

Cryotubes used to store strains of bacteria at low temperature. Bacteria are placed in little holes in the beads inside the tubes and then stored in liquid nitrogen. *(Reproduced by permission of Photo Researchers, Inc.)*

Cryogenics

Cryogens and Their Boiling Points

Cryogen	°F	°C	K
Oxygen	−297	−183	90
Nitrogen	−320	−196	77
Hydrogen	−423	−253	20
Helium	−452	−269	4.2
Neon	−411	−246	27
Argon	−302	−186	87
Krypton	−242	−153	120
Xenon	−161	−107	166

Heat conduction is a relatively simple concept to understand. When two bodies are in contact, heat flows from the body with the higher temperature to the body with a lower temperature. Conduction can occur between any and all forms of matter, whether gas, liquid, or solid. It is essential in the production of cryogenic temperatures and environments. For example, samples may be cooled to cryogenic temperatures by immersing them directly in a cryogenic liquid or by placing them in an atmosphere cooled by cryogenic refrigeration. In either case, the sample cools by conduction (or transfer) of heat to its colder surroundings.

The second process for producing cryogenic conditions is evaporative cooling. Humans are familiar with this process because it is a mechanism by which our bodies lose heat. Atoms and molecules in the gaseous state are moving faster than atoms and molecules in the liquid state. Add heat energy to the particles in a liquid and they will become gaseous. Liquid perspiration on human skin behaves in this way. Perspiration absorbs body heat, becomes a gas, and evaporates from the skin. As a result of that heat loss, the body cools down.

In cryogenics, a container of liquid is allowed to evaporate. Heat from within the liquid is used to convert particles at the surface of the liquid to gas. The gas is then pumped away. More heat from the liquid converts another surface layer of particles to the gaseous state, which is also pumped away. The longer this process continues, the more heat is removed from the liquid and the lower its temperature drops. Once some given temperature is reached, pumping continues at a reduced level in order to maintain the lower temperature. This method can be used to reduce the temperature of any liquid. For example, it can be used to reduce the temperature of liquid nitrogen to its freezing point or to lower the temperature of liquid helium to approximately 1 K.

A third process makes use of the Joule-Thomson effect, which was discovered by English physicist James Prescott Joule (1818–1889) and William Thomson, Lord Kelvin, in 1852. The Joule-Thomson effect depends on the relationship of volume (bulk or mass), pressure, and temperature in a gas. Change any one of these three variables, and at least one of the other two (or both) will also change. Joule and Thomson found, for example, that allowing a gas to expand very rapidly causes its temperature to drop dramatically. Reducing the pressure on a gas accomplishes the same effect.

To cool a gas using the Joule-Thomson effect, the gas is first pumped into a container under high pressure. The container is fitted with a valve with a very small opening. When the valve is opened, the gas escapes from the container and expands quickly. At the same time, its temperature drops. The first great success for the Joule-Thomson effect in cryogenics was achieved by Kamerlingh Onnes in 1908 when he liquefied helium.

The Joule-Thomson effect is an important part of our lives today, even though we may not be aware of it. Ordinary household refrigerators and air conditioners operate on this principle. First, a gas is pressurized and cooled to an intermediate temperature by contact with a colder gas or liquid. Then, the gas is expanded, and its temperature drops still further. The heat needed to keep this cycle operating comes from the inside of the refrigerator or the interior of a room, producing the desired cooling effect.

The fourth process for producing cryogenic temperatures uses a phenomenon known as adiabatic demagnetization. Adiabatic demagnetization makes use of special substances known as paramagnetic salts. A paramagnetic salt consists of a very large collection of particles that act like tiny (atom-sized) magnets. Normally these magnetic particles are spread out in all possible directions. As a result, the salt itself is not magnetic. That condition changes when the salt is placed between the poles of a magnet. The magnetic field of the magnet causes all the tiny magnetic particles in the salt to line up in the same direction. The salt becomes magnetic, too.

At this exact moment, however, suppose that the external magnet is taken away and the paramagnetic salt is placed within a liquid. Almost immediately, the tiny magnetic particles within the salt return to their random, every-which-way condition. To make this change, however, the

Cryogenics

particles require an input of energy. In this example, the energy is taken from the liquid into which the salt was placed. As the liquid gives up energy to the paramagnetic salt, its temperature drops.

Adiabatic demagnetization has been used to produce some of the coldest temperatures ever observed—within a few thousandths of a degree kelvin of absolute zero. A related process involving the magnetization and demagnetization of atomic nuclei is known as nuclear demagnetization. With nuclear demagnetization, temperatures within a few millionths of a degree of absolute zero have been reached.

Applications

Following his successful liquefaction of helium in 1908, Kamerlingh Onnes turned his attention to the study of properties of other materials at very low temperatures. The first property he investigated was the electrical resistance of metals. Electrical resistance is the tendency of a substance to prevent the flow of an electrical current through it. Scientists had long known that electrical resistance tends to decrease with decreasing temperature. They assumed that resistance would completely disappear at absolute zero.

Research in this area had great practical importance. All electrical appliances (ovens, toasters, television sets, and radios, for example) operate with low efficiency because so much energy is wasted in overcoming electrical resistance. An appliance with no electrical resistance could operate at much less cost than existing appliances.

What Onnes discovered, however, was that for some metals, electrical resistance drops to zero very suddenly at temperatures above absolute zero. The effect is called superconductivity and has some very important applications in today's world. For example, superconductors are used to make magnets for particle accelerators (devices used, among other things, to study subatomic particles such as electrons and protons) and for magnetic resonance imaging (MRI) systems (a diagnostic tool used in many hospitals).

The discovery of superconductivity led other scientists to study a variety of material properties at cryogenic temperatures. Today, physicists, chemists, material scientists, and biologists study the properties of metals, as well as the properties of insulators, semiconductors, plastics, composites, and living tissue. Over the years, this research has resulted in the identification of a number of useful properties. One such property common to most materials that are subjected to extremely low temperatures is brittleness. The recycling industry takes advantage of this by immersing recyclables in liquid nitrogen, after which they are easily pulverized and separated for reprocessing.

Still another cryogenic material property that is sometimes useful is that of thermal contraction. Materials shrink when cooled. To a point (about the temperature of liquid nitrogen), the colder a material gets the more it shrinks. An example is the use of liquid nitrogen in the assembly of some automobile engines. In order to get extremely tight fits when installing valve seats, for instance, the seats are cooled to liquid nitrogen temperatures, whereupon they contract and are easily inserted in the engine head. When they warm up, a perfect fit results.

Cryogenic liquids are also used in the space program. For example, cryogenic materials are used to propel rockets into space. A tank of liquid hydrogen provides the fuel to be burned and a second tank of liquid oxygen is provided for combustion.

Another space application of cryogenics is the use of liquid helium to cool orbiting infrared telescopes. Infrared telescopes detect objects in space not from the light they give off but from the infrared radiation (heat) they emit. However, the operation of the telescope itself also gives off heat. What can be done to prevent the instrument from being blinded by its own heat to the infrared radiation from stars? The answer is to cool parts of the telescope with liquid helium. At the temperature of liquid helium (1.8 K) the telescope can easily pick up infrared radiation of the stars, whose temperature is about 3 K.

Crystal

A crystal is a solid whose particles are arranged in an orderly, repeating, geometric pattern. Crystals come in all sizes and shapes. Regular table salt, for instance, consists of tiny cubic particles called salt crystals.

Crystals at the atomic level

The crystal shapes that we can see with our naked eye reflect a similar geometric pattern that exists at the level of atoms. For example, the substance we call salt is actually a chemical compound called sodium chloride. Sodium chloride is made of sodium ions and chloride ions. Each sodium ion is a tiny particle carrying a single positive electric charge. Each chloride ion is a tiny particle carrying a single negative charge.

A crystal of table salt consists of trillions and trillions of sodium ions and chloride ions. The way in which these ions arrange themselves in a

Crystal

salt crystal depends on two factors: the size of each ion and the electric charge on each ion. Recall that similar (or like) electric charges repel each other, and unlike charges attract each other. That fact means that all of the sodium ions tend to repel each other. They will get as far from each other as possible in a salt crystal. The same is true of the chloride ions.

On the other hand, positively charged sodium ions will be attracted by negatively charged chloride ions. Sodium ions and chloride ions will try to get as close to each other as possible. Obviously some kind of compromise position has to be found that allows these forces of attraction and repulsion to be balanced against each other. Size can make a difference, too. Chloride ions are much larger than sodium ions, and this affects the shape of salt crystals.

Unit cells and crystal lattice

The final compromise that best satisfies charge and size factors in sodium chloride is a cube. One sodium ion occurs at each of the alternate corners of the cube. One chloride ion occurs at the other alternate corners of the cube. In this arrangement, sodium ions and chloride ions are held together by forces of electrical attraction, but ions of the same kind are kept as far from each other as possible.

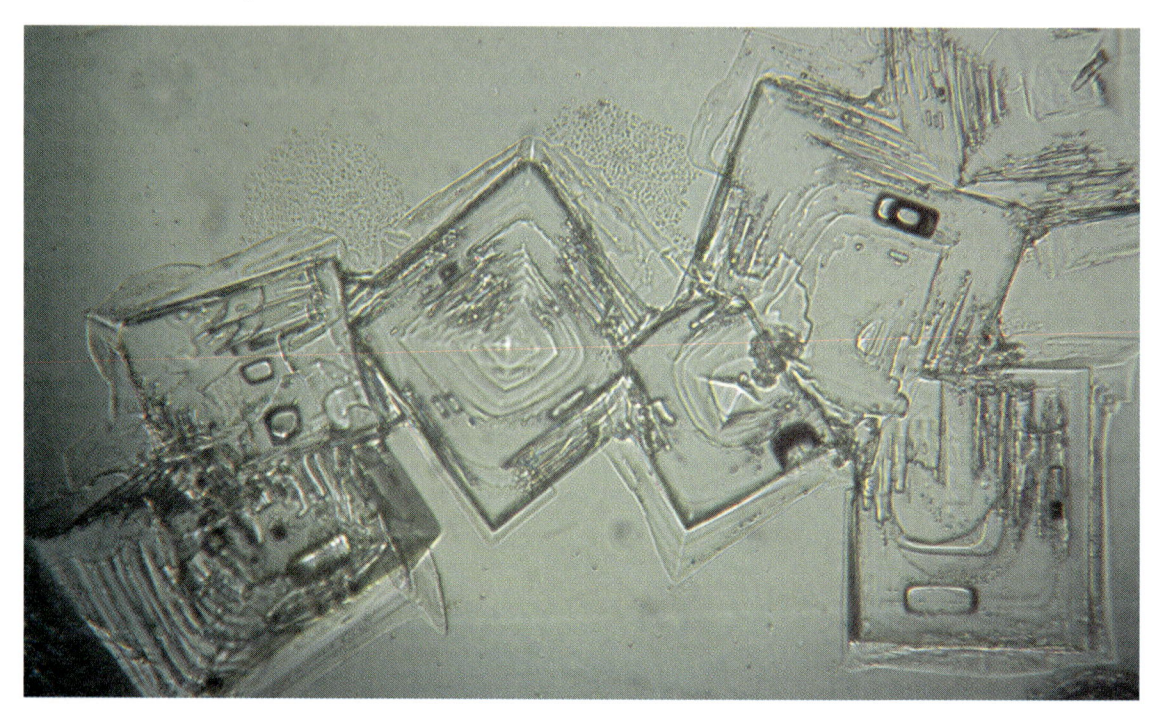

Salt crystals magnified to 40 times their original size, displaying its crystal lattice. (Reproduced by permission of JLM Visuals.)

Crystal

The basic shape that satisfies charge and size factors is known as a unit cell. Thus, for sodium chloride, the unit cell is a cube. Other geometric arrangements are also possible. For example, the problem of balancing electric charges for a crystal of calcium chloride is different than it is for sodium chloride. The calcium ions in calcium chloride each have a positive electric charge of two units. A different geometric arrangement is necessary to accommodate doubly charged calcium ions and singly charged chloride ions.

There are seven geometric shapes that crystals can assume. In a tetragonal crystal system, for example, ions are arranged at the corners of

Sodium chloride (salt) crystals. *(Reproduced by permission of JLM Visuals.)*

Currents, ocean

> **Words to Know**
>
> **Ion:** A molecule or atom that has lost one or more electrons and is, therefore, electrically charged.
>
> **Lattice:** A collection of unit cells that are all identical.
>
> **Unit cell:** The simplest three-dimensional structure of which a crystal is made.

a rectangular box whose end is a square. In an orthorhombic crystal system, ions are arranged at the corners of a rectangular box whose end is a rectangle.

If you could look at a crystal of table salt with a microscope you would see a vast system of unit cells. That system is known as a crystal lattice. As shown in the photo of salt (on p. 602), a crystal lattice is simply many unit cells—all exactly alike—joined to each other. For this reason, the crystal shape that you can actually see for a crystal of table salt reflects exactly what the unit cell for sodium chloride looks like.

Currents, ocean

Currents are steady, smooth movements of water following either a straight or circular path. All of the water in the oceans on Earth are in constant circulation. The two main types of ocean currents are surface currents and deep water or bottom currents.

Surface currents

Surface currents are the most obvious type of current. They are created mainly by prevailing winds, such as trade winds around the equator or westerly winds over the middle latitudes. When wind blows across the water surface, it sets the water in motion. Surface currents can extend to depths of about 65 feet (20 meters).

Landmasses—continental coasts and islands—also affect surface currents. Landmasses act as barriers to the natural path of currents. Without landmasses, there would be a uniform ocean movement from west to east at middle latitudes and from east to west near the equator and at the poles. Instead, landmasses deflect currents or split them up into branches. This deflection, combined with the rotation of Earth on it axis, forces surface currents to flow in circular patterns. These patterns, called gyres (pronounced JEYE-ers), flow clockwise in the Northern Hemisphere and counterclockwise in the Southern Hemisphere.

Surface currents help to moderate Earth's temperatures. As surface currents move, they absorb heat in the tropical regions and release it in colder environments near the poles. The Gulf Stream, a major surface current that originates in the Gulf of Mexico, illustrates this. Traveling at an average of 4 miles (6.4 kilometers) per hour, it carries warm water up the east coast of North America and flows across the Atlantic Ocean, where it warms the climate of England and Northern Europe.

Deep water currents

Deep water currents move very slowly, usually around 1 inch (2.5 centimeters) per second. However, they are responsible for circulating 90 percent of Earth's ocean water. This circulation influences not only weather patterns but the overall health of the oceans.

Deep water currents are set in motion by variations in water density, which is directly related to temperature and salinity, or salt level. Colder, saltier water is heavier than warmer, fresher water. Water near the poles is colder and saltier than water near the equator. This cold water sinks and flows beneath the ocean surface toward the equator, where it is warmed. It then rises to replace the water that surface currents constantly carry toward the poles.

[See also **Ocean; Oceanography; Tides**]

Cybernetics

Cybernetics is the study of communication and control processes in living organisms and machines. Cybernetics analyzes the ability of humans, animals, and some machines to respond to or make adjustments based upon input from the environment. This process of response or adjustment is called feedback or automatic control. Feedback helps people and machines control their actions by telling them whether they are proceeding in the right direction.

Cybernetics

> **Words to Know**
>
> **Artificial intelligence (AI):** The science that attempts to imitate human intelligence with computers.
>
> **Feedback:** Information that tells a system what the results of its actions are.
>
> **Robotics:** The science that deals with the design and construction of robots.

For example, a household thermostat uses feedback when it turns a furnace on or off based on its measurements of temperature. A human being, on the other hand, is such a complex system that the simplest action involves complicated feedback loops. A hand picking up a glass of milk is guided continually by the brain that receives feedback from the eyes and hand. The brain decides in an instant where to grasp the glass and where to raise it in order to avoid collisions and prevent spillage.

The earliest known feedback control mechanism, the centrifugal governor, was developed by the Scottish inventor James Watt in 1788. Watt's steam engine governor kept the engine running at a constant rate.

Systems for guiding missiles

The principles for feedback control were first clearly defined by American mathematician Norbert Wiener (1894–1964). With his colleague Julian Bigelow, Wiener worked for the U.S. government during World War II (1939–45), developing radar and missile guidance systems using automatic information processing and machine controls.

After the war, Wiener continued to work in machine and human feedback research. In 1950, he published *The Human Use of Human Beings: Cybernetics and Society.* In this work, Wiener cautioned that an increased reliance on machines might start a decline in human intellectual capabilities. Wiener also coined the word "cybernetics," which comes from the Greek word *kybernetes,* meaning "steersman."

Cybernetics and industry

With the advent of the digital computer, cybernetic principles such as those described by Wiener were applied to increasingly complex

Cybernetics

tasks. The result was machines with the practical ability to carry out meaningful work. In 1946, Delmar S. Harder devised one of the earliest such systems to automate the manufacture of car engines at the Ford Motor Company. The system involved an element of thinking: the machines regulated themselves, without human supervision, to produce the desired results. Harder's assembly-line automation produced one car engine every 14 minutes, compared with the 21 hours if had previously taken human workers.

By the 1960s and 1970s, the field of cybernetics, robotics, and artificial intelligence began to skyrocket. A large number of industrial and manufacturing plants devised and installed cybernetic systems such as robots in the workplace. In 1980, there were roughly 5,000 industrial robots in the United States. By the early twenty-first century, researchers estimated there were as many as 500,000.

Considerable research is now focused on creating computers that imitate the workings of the human mind. The eventual aim, and the continuing area of research in this field, is the production of a neural computer, in which the architecture of the human brain is reproduced. The system would be brought about by transistors and resistors acting such as neurons, axons, and dendrites do in the brain. The advantage of neural computers is they will be able to grow and adapt. They will be able to

Hannibal (Attila II), developed at the Massachusetts Institute of Technology's Artificial Intelligence Lab: a machine that "thinks." *(Reproduced by permission of Photo Researchers, Inc.)*

Cyclone and anticyclone

learn from past experience and recognize patterns. This will enable them to operate intuitively, at a faster rate, and in a predictable manner.

[*See also* **Artificial intelligence; Computer, digital; Robotics**]

Cyclamate

Cyclamate is the name given to a family of organic compounds that became popular in the 1950s as artificial sweeteners. They are about 30 times as sweet as ordinary table sugar (sucrose) but have none of sugar's calories. By the mid-1960s, a combination of cyclamate and saccharin (another artificial sweetener) known as Sucaryl had become one of the most popular alternatives to sugar.

Trouble arose in 1969, however. A scientific study showed that among rats fed a high dose of Sucaryl for virtually their whole lives, about 15 percent developed bladder cancer. Presented with this information, the U.S. Food and Drug Administration (FDA) decided to ban the use of cyclamate in foods. In 1973, Abbott Laboratories, the makers of cyclamate, petitioned the FDA to change its mind and allow the use of cyclamates once more. Abbott presented a number of studies showing that cyclamate does not cause bladder cancer in rats or have other harmful health effects.

The FDA studied Abbott's petition for seven years before deciding to reject it. Cyclamate remained banned for use in foods. In 1982, Abbott submitted a second petition asking for approval of cyclamate. As of the beginning of 2001, the FDA had not acted on that petition.

What makes this case of special interest is that a potentially important food additive has been banned on the basis of a single scientific study. More than two dozen other studies on its safety reportedly failed to show the same results. Furthermore, the second component of Sucaryl—saccharin—has also been shown to cause bladder cancer in experimental animals. Yet the FDA continues to allow its use in foods.

Cyclone and anticyclone

A cyclone is a storm or system of winds that rotates around a center of low atmospheric pressure. An anticyclone is a system of winds that rotates around a center of high atmospheric pressure. Distinctive weather patterns tend to be associated with both cyclones and anticyclones. Cyclones (commonly known as lows) generally are indicators of rain, clouds, and other forms of bad weather. Anticyclones (commonly known as highs) are predictors of fair weather.

Winds in a cyclone blow counterclockwise in the Northern Hemisphere and clockwise in the Southern Hemisphere. Winds in an anticyclone blow just the opposite. Vertical air movements are associated with both cyclones and anticyclones. In cyclones, air close to the ground is forced inward toward the center of the cyclone, where pressure is lowest. It then begins to rise upward, expanding and cooling in the process. This cooling increases the humidity of the rising air, which results in cloudiness and high humidity in the cyclone.

In anticyclones, the situation is reversed. Air at the center of an anticyclone is forced away from the high pressure that occurs there. That air is replaced in the center by a downward draft of air from higher altitudes. As this air moves downward, it is compressed and warmed. This

Cyclone and anticyclone

A computer-enhanced image of Hurricane Diana at its strongest on September 11, 1984. The hurricane was just off the coast of South and North Carolina at the time, and winds within it were 130 miles per hour (210 kilometers per hour). *(Reproduced by permission of National Aeronautics and Space Administration.)*

Cyclone and anticyclone

warming reduces the humidity of the descending air, which results in few clouds and low humidity in the anticyclone.

Hurricanes and typhoons

Cyclones that form over warm tropical oceans are called tropical cyclones (they are also known as tropical storms or tropical depressions). Tropical cyclones usually move toward the west with the flow of trade winds. A tropical cyclone that drastically increases in intensity is known as a hurricane when it occurs in the Atlantic Ocean or adjacent seas. To be classified as a hurricane, a tropical cyclone must produce winds over 74 miles (119 kilometers) per hour. Hurricanes usually generate off the coast of West Africa and move westward toward Central America and the eastern United States. They increase in size and strength until they reach land or more northern latitudes. In addition to high, sustained winds, hurricanes deliver heavy rain and devastating ocean waves.

In the western Pacific Ocean and adjacent seas, a hurricane is known as a typhoon. This word comes from the Cantonese *tai-fung,* meaning "great wind."

[*See also* **Atmospheric pressure; Storm surge; Tornado; Weather; Wind**]

Where to Learn More

Books

Earth Sciences

Cox, Reg, and Neil Morris. *The Natural World.* Philadelphia, PA: Chelsea House, 2000.

Dasch, E. Julius, editor. *Earth Sciences for Students.* Four volumes. New York: Macmillan Reference, 1999.

Denecke, Edward J., Jr. *Let's Review: Earth Science.* Second edition. Hauppauge, NY: Barron's, 2001.

Engelbert, Phillis. *Dangerous Planet: The Science of Natural Disasters.* Three volumes. Farmington Hills, MI: UXL, 2001.

Gardner, Robert. *Human Evolution.* New York: Franklin Watts, 1999.

Hall, Stephen. *Exploring the Oceans.* Milwaukee, WI: Gareth Stevens, 2000.

Knapp, Brian. *Earth Science: Discovering the Secrets of the Earth.* Eight volumes. Danbury, CT: Grolier Educational, 2000.

Llewellyn, Claire. *Our Planet Earth.* New York: Scholastic Reference, 1997.

Moloney, Norah. *The Young Oxford Book of Archaeology.* New York: Oxford University Press, 1997.

Nardo, Don. *Origin of Species: Darwin's Theory of Evolution.* San Diego, CA: Lucent Books, 2001.

Silverstein, Alvin, Virginia Silverstein, and Laura Silverstein Nunn. *Weather and Climate.* Brookfield, CN: Twenty-First Century Books, 1998.

Williams, Bob, Bob Ashley, Larry Underwood, and Jack Herschbach. *Geography.* Parsippany, NJ: Dale Seymour Publications, 1997.

Life Sciences

Barrett, Paul M. *National Geographic Dinosaurs.* Washington, D.C.: National Geographic Society, 2001.

Fullick, Ann. *The Living World.* Des Plaines, IL: Heinemann Library, 1999.

Gamlin, Linda. *Eyewitness: Evolution.* New York: Dorling Kindersley, 2000.

Greenaway, Theresa. *The Plant Kingdom: A Guide to Plant Classification and Biodiversity.* Austin, TX: Raintree Steck-Vaughn, 2000.

Kidd, J. S., and Renee A Kidd. *Life Lines: The Story of the New Genetics.* New York: Facts on File, 1999.

Kinney, Karin, editor. *Our Environment.* Alexandria, VA: Time-Life Books, 2000.

Where to Learn More

Nagel, Rob. *Body by Design: From the Digestive System to the Skeleton.* Two volumes. Farmington Hills, MI: UXL., 2000.

Parker, Steve. *The Beginner's Guide to Animal Autopsy: A "Hands-in" Approach to Zoology, the World of Creatures and What's Inside Them.* Brookfield, CN: Copper Beech Books, 1997.

Pringle, Laurence. *Global Warming: The Threat of Earth's Changing Climate.* New York: SeaStar Books, 2001.

Riley, Peter. *Plant Life.* New York: Franklin Watts, 1999.

Stanley, Debbie. *Genetic Engineering: The Cloning Debate.* New York: Rosen Publishing Group, 2000.

Whyman, Kate. *The Animal Kingdom: A Guide to Vertebrate Classification and Biodiversity.* Austin, TX: Raintree Steck-Vaughn, 1999.

Physical Sciences

Allen, Jerry, and Georgiana Allen. *The Horse and the Iron Ball: A Journey Through Time, Space, and Technology.* Minneapolis, MN: Lerner Publications, 2000.

Berger, Samantha, *Light.* New York: Scholastic, 1999.

Bonnet, Bob L., and Dan Keen. *Physics.* New York: Sterling Publishing, 1999.

Clark, Stuart. *Discovering the Universe.* Milwaukee, WI: Gareth Stevens, 2000.

Fleisher, Paul, and Tim Seeley. *Matter and Energy: Basic Principles of Matter and Thermodynamics.* Minneapolis, MN: Lerner Publishing, 2001.

Gribbin, John. *Eyewitness: Time and Space.* New York: Dorling Kindersley, 2000.

Holland, Simon. *Space.* New York: Dorling Kindersley, 2001.

Kidd, J. S., and Renee A. Kidd. *Quarks and Sparks: The Story of Nuclear Power.* New York: Facts on File, 1999.

Levine, Shar, and Leslie Johnstone. *The Science of Sound and Music.* New York: Sterling Publishing, 2000

Naeye, Robert. *Signals from Space: The Chandra X-ray Observatory.* Austin, TX: Raintree Steck-Vaughn, 2001.

Newmark, Ann. *Chemistry.* New York: Dorling Kindersley, 1999.

Oxlade, Chris. *Acids and Bases.* Chicago, IL: Heinemann Library, 2001.

Vogt, Gregory L. *Deep Space Astronomy.* Brookfield, CT: Twenty-First Century Books, 1999.

Technology and Engineering Sciences

Baker, Christopher W. *Scientific Visualization: The New Eyes of Science.* Brookfield, CT: Millbrook Press, 2000.

Cobb, Allan B. *Scientifically Engineered Foods: The Debate over What's on Your Plate.* New York: Rosen Publishing Group, 2000.

Cole, Michael D. *Space Launch Disaster: When Liftoff Goes Wrong.* Springfield, NJ: Enslow, 2000.

Deedrick, Tami. *The Internet.* Austin, TX: Raintree Steck-Vaughn, 2001.

DuTemple, Leslie A. *Oil Spills.* San Diego, CA: Lucent Books, 1999.

Gaines, Ann Graham. *Satellite Communication.* Mankata, MN: Smart Apple Media, 2000.

Gardner, Robert, and Dennis Shortelle. *From Talking Drums to the Internet: An Encyclopedia of Communications Technology.* Santa Barbara, CA: ABC-Clio, 1997.

Graham, Ian S. *Radio and Television.* Austin, TX: Raintree Steck-Vaughn, 2000.

Parker, Steve. *Lasers: Now and into the Future.* Englewood Cliffs, NJ: Silver Burdett Press, 1998.

Sachs, Jessica Snyder. *The Encyclopedia of Inventions.* New York: Franklin Watts, 2001.

Wilkinson, Philip. *Building.* New York: Dorling Kindersley, 2000.

Wilson, Anthony. *Communications: How the Future Began.* New York: Larousse Kingfisher Chambers, 1999.

Periodicals

Archaeology. Published by Archaeological Institute of America, 656 Beacon Street, 4th Floor, Boston, Massachusetts 02215. Also online at www.archaeology.org.

Astronomy. Published by Kalmbach Publishing Company, 21027 Crossroads Circle, Brookfield, WI 53186. Also online at www.astronomy.com.

Discover. Published by Walt Disney Magazine, Publishing Group, 500 S. Buena Vista, Burbank, CA 91521. Also online at www.discover.com.

National Geographic. Published by National Geographic Society, 17th & M Streets, NW, Washington, DC 20036. Also online at www.nationalgeographic.com.

New Scientist. Published by New Scientist, 151 Wardour St., London, England W1F 8WE. Also online at www.newscientist.com (includes links to more than 1,600 science sites).

Popular Science. Published by Times Mirror Magazines, Inc., 2 Park Ave., New York, NY 10024. Also online at www.popsci.com.

Science. Published by American Association for the Advancement of Science, 1333 H Street, NW, Washington, DC 20005. Also online at www.sciencemag.org.

Science News. Published by Science Service, Inc., 1719 N Street, NW, Washington, DC 20036. Also online at www.sciencenews.org.

Scientific American. Published by Scientific American, Inc., 415 Madison Ave, New York, NY 10017. Also online at www.sciam.com.

Smithsonian. Published by Smithsonian Institution, Arts & Industries Bldg., 900 Jefferson Dr., Washington, DC 20560. Also online at www.smithsonianmag.com.

Weatherwise. Published by Heldref Publications, 1319 Eighteenth St., NW, Washington, DC 20036. Also online at www.weatherwise.org.

Web Sites

Cyber Anatomy (provides detailed information on eleven body systems and the special senses) *http://library.thinkquest.org/11965/*

The DNA Learning Center (provides in-depth information about genes for students and educators) *http://vector.cshl.org/*

Educational Hotlists at the Franklin Institute (provides extensive links and other resources on science subjects ranging from animals to wind energy) *http://sln.fi.edu/tfi/hotlists/hotlists.html*

ENC Web Links: Science (provides an extensive list of links to sites covering subject areas under earth and space science, physical science, life science, process skills, and the history of science) *http://www.enc.org/weblinks/science/*

ENC Web Links: Math topics (provides an extensive list of links to sites covering subject areas under topics such as advanced mathematics, algebra, geometry, data analysis and probability, applied mathematics, numbers and operations, measurement, and problem solving) *http://www.enc.org/weblinks/math/*

Encyclopaedia Britannica Discovering Dinosaurs Activity Guide *http://dinosaurs.eb.com/dinosaurs/study/*

The Exploratorium: The Museum of Science, Art, and Human Perception *http://www.exploratorium.edu/*

Where to Learn More

Where to Learn More

ExploreMath.com (provides highly interactive math activities for students and educators) http://www.exploremath.com/

ExploreScience.com (provides highly interactive science activities for students and educators) http://www.explorescience.com/

Imagine the Universe! (provides information about the universe for students aged 14 and up) http://imagine.gsfc.nasa.gov/

Mad Sci Network (highly searchable site provides extensive science information in addition to a search engine and a library to find science resources on the Internet; also allows students to submit questions to scientists) http://www.madsci.org/

The Math Forum (provides math-related information and resources for elementary through graduate-level students) http://forum.swarthmore.edu/

NASA Human Spaceflight: International Space Station (NASA homepage for the space station) http://www.spaceflight.nasa.gov/station/

NASA's Origins Program (provides up-to-the-minute information on the scientific quest to understand life and its place in the universe) http://origins.jpl.nasa.gov/

National Human Genome Research Institute (provides extensive information about the Human Genome Project) http://www.nhgri.nih.gov:80/index.html

New Scientist Online Magazine http://www.newscientist.com/

The Nine Planets (provides a multimedia tour of the history, mythology, and current scientific knowledge of each of the planets and moons in our solar system) http://seds.lpl.arizona.edu/nineplanets/nineplanets/nineplanets.html

The Particle Adventure (provides an interactive tour of quarks, neutrinos, antimatter, extra dimensions, dark matter, accelerators, and particle detectors) http://particleadventure.org/

PhysLink: Physics and astronomy online education and reference http://physlink.com/

Savage Earth Online (online version of the PBS series exploring earthquakes, volcanoes, tsunamis, and other seismic activity) http://www.pbs.org/wnet/savageearth/

Science at NASA (provides breaking information on astronomy, space science, earth science, and biological and physical sciences) http://science.msfc.nasa.gov/

Science Learning Network (provides Internet-guided science applications as well as many middle school science links) http://www.sln.org/

SciTech Daily Review (provides breaking science news and links to dozens of science and technology publications; also provides links to numerous "interesting" science sites) http://www.scitechdaily.com/

Space.com (space news, games, entertainment, and science fiction) http://www.space.com/index.html

SpaceDaily.com (provides latest news about space and space travel) http://www.spacedaily.com/

SpaceWeather.com (science news and information about the Sun-Earth environment) http://www.spaceweather.com/

The Why Files (exploration of the science behind the news; funded by the National Science Foundation) http://whyfiles.org/

Index

Italic type indicates volume numbers; **boldface** type indicates entries and their page numbers; (ill.) indicates illustrations.

A

Abacus *1:* **1-2** 1 (ill.)
Abelson, Philip *1:* 24
Abortion *3:* 565
Abrasives *1:* **2-4,** 3 (ill.)
Absolute dating *4:* 616
Absolute zero *3:* 595-596
Abyssal plains *7:* 1411
Acceleration *1:* **4-6**
Acetylsalicylic acid *1:* **6-9,** 8 (ill.)
Acheson, Edward G. *1:* 2
Acid rain *1:* **9-14,** 10 (ill.), 12 (ill.), *6:* 1163, *8:* 1553
Acidifying agents *1:* 66
Acids and bases *1:* **14-16,** *8:* 1495
Acoustics *1:* **17-23,** 17 (ill.), 20 (ill.)
Acquired immunodeficiency syndrome. *See* **AIDS (acquired immunodeficiency syndrome)**
Acrophobia *8:* 1497
Actinides *1:* **23-26,** 24 (ill.)
Acupressure *1:* 121
Acupuncture *1:* 121
Adams, John Couch *7:* 1330
Adaptation *1:* **26-32,** 29 (ill.), 30 (ill.)
Addiction *1:* **32-37,** 35 (ill.), *3:* 478
Addison's disease *5:* 801
Adena burial mounds *7:* 1300
Adenosine triphosphate *7:* 1258
ADHD *2:* 237-238
Adhesives *1:* **37-39,** 38 (ill.)
Adiabatic demagnetization *3:* 597
ADP *7:* 1258
Adrenal glands *5:* 796 (ill.)
Adrenaline *5:* 800
Aerobic respiration *9:* 1673
Aerodynamics *1:* **39-43,** 40 (ill.)
Aerosols *1:* **43-49,** 43 (ill.)
Africa *1:* **49-54,** 50 (ill.), 53 (ill.)
Afterburners *6:* 1146
Agent Orange *1:* **54-59,** 57 (ill.)
Aging and death *1:* **59-62**
Agoraphobia *8:* 1497
Agriculture *1:* **62-65,** 63, 64 (ill.), *3:*582-590, *5:* 902-903, *9:* 1743-744, *7:* 1433 (ill.)
Agrochemicals *1:* **65-69,** 67 (ill.), 68 (ill.)
Agroecosystems *2:* 302
AI. *See* **Artificial intelligence**
AIDS (acquired immunodeficiency syndrome) *1:* **70-74,** 72 (ill.), *8:* 1583, *9:* 1737
Air flow *1:* 40 (ill.)
Air masses and fronts *1:* **80-82,** 80 (ill.)
Air pollution *8:* 1552, 1558
Aircraft *1:* **74-79,** 75 (ill.), 78 (ill.)
Airfoil *1:* 41
Airplanes. *See* **Aircraft**
Airships *1:* 75

Index

Al-jabr wa'l Muqabalah 1: 97
Al-Khwarizmi *1:* 97
Alchemy *1:* **82-85**
Alcohol (liquor) *1:* 32, 85-87
Alcoholism *1:* **85-88**
Alcohols *1:* **88-91,** 89 (ill.)
Aldrin, Edwin *9:* 1779
Ale *2:* 354
Algae *1:* **91-97,** 93 (ill.), 94 (ill.)
Algal blooms *1:* 96
Algebra *1:* **97-99,** *2:* 333-334
Algorithms *1:* 190
Alkali metals *1:* **99-102,** 101 (ill.)
Alkaline earth metals *1:* **102-106,** 104 (ill.)
Alleles *7:* 1248
Allergic rhinitis *1:* 106
Allergy *1:* **106-110,** 108 (ill.)
Alloy *1:* **110-111**
Alpha particles *2:* 233, *8:* 1620, 1632
Alps *5:* 827, *7:* 1301
Alternating current (AC) *4:* 741
Alternation of generations *9:* 1667
Alternative energy sources *1:* **111-118,** 114 (ill.), 115 (ill.), *6:* 1069
Alternative medicine *1:* **118-122**
Altimeter *2:* 266
Aluminum *1:* 122-124, 125 (ill.)
Aluminum family *1:* **122-126,** 125 (ill.)
Alzheimer, Alois *1:* 127
Alzheimer's disease *1:* 62, **126-130,** 128 (ill.)
Amazon basin *9:* 1774
American Red Cross *2:* 330
Ames test *2:* 408
Amino acid *1:* **130-131**
Aminoglycosides *1:* 158
Ammonia *7:* 1346
Ammonification *7:* 1343
Amniocentesis *2:* 322
Amoeba *1:* **131-134,** 132 (ill.)
Ampere *3:* 582, *4:* 737
Amère, André *4:* 737, *6:* 1212
Ampere's law *4:* 747
Amphibians *1:* **134-137,** 136 (ill.)
Amphiboles *1:* 191
Amphineura *7:* 1289
Amplitude modulation *8:* 1627
Amundsen, Roald *1:* 152
Anabolism *7:* 1255

Anaerobic respiration *9:* 1676
Anatomy *1:* **138-141,** 140 (ill.)
Anderson, Carl *1:* 163, *4:* 773
Andes Mountains *7:* 1301, *9:* 1775-1776
Andromeda galaxy *5:* 939 (ill.)
Anemia *1:* 8, *6:* 1220
Aneroid barometer *2:* 266
Anesthesia *1:* **142-145,** 143 (ill.)
Angel Falls *9:* 1774
Angiosperms *9:* 1729
Animal behavior *2:* 272
Animal hormones *6:* 1053
Animal husbandry *7:* 1433
Animals *1:* **145-147,** 146 (ill.), *6:* 1133-1134
Anorexia nervosa *4:* 712
Antarctic Treaty *1:* 153
Antarctica *1:* **147-153,** 148 (ill.), 152 (ill.)
Antennas *1:* **153-155,** 154 (ill.)
Anthrax *2:* 287
Antibiotics *1:* **155-159,** 157 (ill.)
Antibody and antigen *1:* **159-162,** *2:* 311
Anticyclones, cyclones and *3:* 608-610
Antidiuretic hormone *5:* 798
Antigens, antibodies and *1:* 159-162
Antimatter *1:* 163
Antimony *7:* 1348
Antiparticles *1:* **163-164**
Antiprotons *1:* 163
Antipsychotic drugs *8:* 1598
Antiseptics *1:* **164-166**
Anurans *1:* 136
Apennines *5:* 827
Apes *8:* 1572
Apgar Score *2:* 322
Aphasia *9:* 1798, 1799
Apollo 11 *9:* 1779, 1780 (ill.)
Apollo objects *1:* 202
Appalachian Mountains *7:* 1356
Appendicular skeleton *9:* 1741
Aquaculture *1:* **166-168,** 167 (ill.)
Arabian Peninsula. See **Middle East**
Arabic numbers. See **Hindu-Arabic number system**
Arachnids *1:* **168-171,** 170 (ill.)
Arachnoid *2:* 342
Arachnophobia *8:* 1497
Ararat, Mount *1:* 197

Archaeoastronomy *1:* **171-173,** 172 (ill.)
Archaeology *1:* **173-177,** 175 (ill.), 176 (ill.), *7:* 1323-1327
Archaeology, oceanic. *See* **Nautical archaeology**
Archaeopteryx lithographica *2:* *312*
Archimedes *2:* *360*
Archimedes' Principle *2:* *360*
Argon *7:* 1349, 1350
Ariel *10:* 1954
Aristotle *1:* 138, *2:* 291, *5:* 1012, *6:* 1169
Arithmetic *1:* 97, **177-181,** *3:* 534-536
Arkwright, Edmund *6:* 1098
Armstrong, Neil *9:* 1779
Arnold of Villanova *2:* 404
ARPANET *6:* 1124
Arrhenius, Svante *1:* 14, *8:* 1495
Arsenic *7:* 1348
Arthritis *1:* **181-183,** 182 (ill.)
Arthropods *1:* **183-186,** 184 (ill.)
Artificial blood *2:* 330
Artificial fibers *1:* **186-188,** 187 (ill.)
Artificial intelligence *1:* **188-190,** *2:* 244
Asbestos *1:* **191-194,** 192 (ill.), *6:* 1092
Ascorbic acid. *See* **Vitamin C**
Asexual reproduction *9:* 1664 (ill.), 1665
Asia *1:* **194-200,** 195 (ill.), 198 (ill.)
Aspirin. *See* **Acetylsalicylic acid**
Assembly language *3:* 551
Assembly line *7:* 1238
Astatine *6:* 1035
Asterisms *3:* 560
Asteroid belt *1:* 201
Asteroids *1:* **200-204,** 203 (ill.), *9:* 1764
Asthenosphere *8:* 1535, 1536
Asthma *1:* **204-207,** 206 (ill.), *9:* 1681
Aston, William *7:* 1240
Astronomia nova *3:* *425*
Astronomy, infrared *6:* 1100-1103
Astronomy, ultraviolet *10:* 1943-1946
Astronomy, x-ray *10:* 2038-2041
Astrophysics *1:* **207-209,** 208 (ill.)
Atherosclerosis *3:* 484
Atmosphere observation *2:* **215-217,** 216 (ill.)

Atmosphere, composition and structure *2:* **211-215,** 214 (ill.)
Atmospheric circulation *2:* **218-221,** 220 (ill.)
Atmospheric optical effects *2:* **221-225,** 223 (ill.)
Atmospheric pressure *2:* **225,** 265, *8:* 1571
Atom *2:* **226-229,** 227 (ill.)
Atomic bomb *7:* 1364, 1381
Atomic clocks *10:* 1895-1896
Atomic mass *2:* 228, **229-232**
Atomic number *4:* 777
Atomic theory *2:* **232-236,** 234 (ill.)
ATP *7:* 1258
Attention-deficit hyperactivity disorder (ADHD) *2:* **237-238**
Audiocassettes. *See* **Magnetic recording/audiocassettes**
Auer metal *6:* 1165
Auroras *2:* 223, 223 (ill.)
Australia *2:* **238-242,** 239 (ill.), 241 (ill.)
Australopithecus afarensis *6:* 1056, 1057 (ill.)
Australopithecus africanus *6:* 1056
Autistic savants. *See* **Savants**
Autoimmune diseases *1:* 162
Automation *2:* **242-245,** 244 (ill.)
Automobiles *2:* **245-251,** 246 (ill.), 249 (ill.)
Autosomal dominant disorders *5:* 966
Auxins *6:* 1051
Avogadro, Amadeo *7:* 1282
Avogadro's number *7:* 1282
Axial skeleton *9:* 1740
Axioms *1:* 179
Axle *6:* 1207
Ayers Rock *2:* 240
AZT *1:* 73

B

B-2 Stealth Bomber *1:* 78 (ill.)
Babbage, Charles *3:* 547
Babbitt, Seward *9:* 1691
Bacitracin *1:* 158
Bacteria *2:* **253-260,** 255 (ill.), 256 (ill.), 259 (ill.)
Bacteriophages *10:* 1974

Index

Baekeland, Leo H.   *8:* 1565
Bakelite   *8:* 1565
Balard, Antoine   *6:* 1034
Baldwin, Frank Stephen   *2:* 371
Ballistics   *2:* **260-261**
Balloons   *1:* 75, *2:* **261-265**, 263 (ill.), 264 (ill.)
Bardeen, John   *10:* 1910
Barite   *6:* 1093
Barium   *1:* 105
Barnard, Christiaan   *6:* 1043, *10:* 1926
Barometer   *2:* **265-267**, 267 (ill.)
Barrier islands   *3:* 500
Bases, acids and 1: 14-16
Basophils   *2:* 329
Bats   *4:* 721
Battery   *2:* **268-270**, 268 (ill.)
Battle fatigue   *9:* 1826
Beaches, coasts and   *3:* 498-500
Becquerel, Henri   *8:* 1630
Bednorz, Georg   *10:* 1851
Behavior (human and animal), study of. See **Psychology**
Behavior   *2:* **270-273**, 271 (ill.), 272 (ill.)
Behaviorism (psychology)   *8:* 1595
Bell Burnell, Jocelyn   *7:* 1340
Bell, Alexander Graham   *10:* 1867 (ill.)
Benthic zone   *7:* 1415
Benz, Karl Friedrich   *2:* 246 (ill.)
Berger, Hans   *9:* 1745
Beriberi   *6:* 1219, *10:* 1982
Bernoulli's principle   *1:* 40, 42, *5:* 884
Beryllium   *1:* 103
Berzelius, Jöns Jakob   *2:* 230
Bessemer converter   *7:* 1445, *10:* 1916
Bessemer, Henry   *10:* 1916
Beta carotene   *10:* 1984
Beta particles   *8:* 1632
Bichat, Xavier   *1:* 141
Big bang theory   *2:* **273-276**, 274 (ill.), *4:* 780
Bigelow, Julian   *3:* 606
Binary number system   *7:* 1397
Binary stars   *2:* **276-278**, 278 (ill.)
Binomial nomenclature   *2:* 337
Biochemistry   *2:* **279-280**
Biodegradable   *2:* **280-281**
Biodiversity   *2:* **281-283**, 282 (ill.)
Bioenergy   *1:* 117, *2:* **284-287**, 284 (ill.)
Bioenergy fuels   *2:* 286
Biofeedback   *1:* 119
Biological warfare   *2:* **287-290**
Biological Weapons Convention Treaty   *2:* 290
Biology   *2:* **290-293**, *7:* 1283-1285
Bioluminescence   *6:* 1198
Biomass energy.   See **Bioenergy**
Biomes   *2:* **293-302**, 295 (ill.), 297 (ill.), 301 (ill.)
Biophysics   *2:* **302-304**
Bioremediation   *7:* 1423
Biosphere 2 Project   *2:* 307-309
Biospheres   *2:* **304-309**, 306 (ill.)
Biot, Jean-Baptiste   *7:* 1262
Biotechnology   *2:* **309-312**, 311 (ill.)
Bipolar disorder   *4:* 633
Birds   *2:* **312-315**, 314 (ill.)
Birth   *2:* **315-319**, 317 (ill.), 318 (ill.)
Birth control.   See **Contraception**
Birth defects   *2:* **319-322**, 321 (ill.)
Bismuth   *7:* 1349
Bjerknes, Jacob   *1:* 80, *10:* 2022
Bjerknes, Vilhelm   *1:* 80, *10:* 2022
Black Death   *8:* 1520
Black dwarf   *10:* 2028
Black holes   *2:* **322-326**, 325 (ill.), *9:* 1654
Blanc, Mont   *5:* 827
Bleuler, Eugen   *9:* 1718
Blood   *2:* **326-330**, 328 (ill.), 330, *3:* 483
Blood banks   *2:* 330
Blood pressure   *3:* 483
Blood supply   *2:* **330-333**
Blood vessels   *3:* 482
Blue stars   *9:* 1802
Bode, Johann   *1:* 201
Bode's Law   *1:* 201
Bogs   *10:* 2025
Bohr, Niels   *2:* 235
Bones.   See **Skeletal system**
Bones, study of diseases of or injuries to.   See **Orthopedics**
Boole, George   *2:* 333
Boolean algebra   *2:* **333-334**
Bopp, Thomas   *3:* 529
Borax   *1:* 126, *6:* 1094
Boreal coniferous forests   *2:* 294
Bores, Leo   *8:* 1617
Boron   *1:* 124-126

Boron compounds *6:* 1094
Bort, Léon Teisserenc de *10:* 2021
Bosons *10:* 1831
Botany *2:* **334-337,** 336 (ill.)
Botulism *2:* 258, 288
Boundary layer effects *5:* 885
Bovine growth hormone *7:* 1434
Boyle, Robert *4:* 780
Boyle's law *5:* 960
Braham, R. R. *10:* 2022
Brahe, Tycho *3:* 574
Brain *2:* **337-351,** 339 (ill.), 341 (ill.)
Brain disorders *2:* 345
Brass *10:* 1920
Brattain, Walter *10:* 1910
Breathing *9:* 1680
Brewing *2:* **352-354,** 352 (ill.)
Bridges *2:* **354-358,** 357 (ill.)
Bright nebulae *7:* 1328
British system of measurement *10:* 1948
Bromine *6:* 1034
Bronchitis *9:* 1681
Bronchodilators *1:* 205
Brønsted, J. N. *1:* 15
Brønsted, J. N. *1:* 15
Bronze *2:* 401
Bronze Age *6:* 1036
Brown algae *1:* 95
Brown dwarf *2:* **358-359**
Brucellosis *2:* 288
Bryan, Kirk *8:* 1457
Bubonic plague *8:* 1518
Buckminsterfullerene *2:* 398, 399 (ill.)
Bugs. *See* **Insects**
Bulimia *4:* 1714-1716
Buoyancy *1:* 74, *2:* **360-361,** 360 (ill.)
Burial mounds *7:* 1298
Burns *2:* **361-364,** 362 (ill.)
Bushnell, David *10:* 1834
Butterflies *2:* **364-367,** 364 (ill.)
Byers, Horace *10:* 2022

C

C-12 *2:* 231
C-14 *1:* 176, *4:* 617
Cable television *10:* 1877
Cactus *4:* 635 (ill.)
CAD/CAM *2:* **369-370,** 369 (ill.)

Caffeine *1:* 34
Caisson *2:* 356
Calcite *3:* 422
Calcium *1:* 104 (ill.), 105
Calcium carbonate *1:* 104 (ill.)
Calculators *2:* **370-371,** 370 (ill.)
Calculus *2:* **371-372**
Calderas *6:* 1161
Calendars *2:* **372-375,** 374 (ill.)
Callisto *6:* 1148, 1149
Calories *2:* **375-376,** *6:* 1045
Calving (icebergs) *6:* 1078, 1079 (ill.)
Cambium *10:* 1927
Cambrian period *8:* 1461
Cameroon, Mount *1:* 51
Canadian Shield *7:* 1355
Canals *2:* **376-379,** 378 (ill.)
Cancer *2:* **379-382,** 379 (ill.), 381 (ill.), *10:* 1935
Canines *2:* **382-387,** 383 (ill.), 385 (ill.)
Cannabis sativa *6:* 1224, 1226 (ill.)
Cannon, W. B. *8:* 1516
Capacitor *4:* 749
Carbohydrates *2:* **387-389,** *7:* 1400
Carbon *2:* 396
Carbon compounds, study of. *See* **Organic chemistry**
Carbon cycle *2:* **389-393,** 391 (ill.)
Carbon dioxide *2:* **393-395,** 394 (ill.)
Carbon family *2:* **395-403,** 396 (ill.), 397 (ill.), 399 (ill.)
Carbon monoxide *2:* **403-406**
Carbon-12 *2:* 231
Carbon-14 *4:* 617
Carboniferous period *8:* 1462
Carborundum *1:* 2
Carcinogens *2:* **406-408**
Carcinomas *2:* 381
Cardano, Girolamo *8:* 1576
Cardiac muscle *7:* 1312
Cardiovascular system *3:* 480
Caries *4:* 628
Carlson, Chester *8:* 1502, 1501 (ill.)
Carnot, Nicholas *6:* 1118
Carnot, Sadi *10:* 1885
Carothers, Wallace *1:* 186
Carpal tunnel syndrome *2:* **408-410**
Cartography *2:* **410-412,** 411 (ill.)
Cascade Mountains *7:* 1358
Caspian Sea *5:* 823, 824

Index

Cassini division *9:* 1711
Cassini, Giovanni Domenico *9:* 1711
Cassini orbiter *9:* 1712
CAT scans *2:* 304, *8:* 1640
Catabolism *7:* 1255
Catalysts and catalysis *3:* **413-415**
Catastrophism *3:* **415**
Cathode *3:* **415-416**
Cathode-ray tube *3:* **417-420,** 418 (ill.)
Cats. *See* **Felines**
Caucasus Mountains *5:* 823
Cavendish, Henry *6:* 1069, *7:* 1345
Caves *3:* **420-423,** 422 (ill.)
Cavities (dental) *4:* 628
Cayley, George *1:* 77
CDC *6:* 1180
CDs. *See* **Compact disc**
Celestial mechanics *3:* **423-428,** 427 (ill.)
Cell wall (plants) *3:* 436
Cells *3:* **428-436,** 432 (ill.), 435 (ill.)
Cells, electrochemical *3:* **436-439**
Cellular metabolism *7:* 1258
Cellular/digital technology *3:* **439-441**
Cellulose *2:* 389, *3:* **442-445,** 442 (ill.)
Celsius temperature scale *10:* 1882
Celsius, Anders *10:* 1882
Cenozoic era *5:* 990, *8:* 1462
Center for Disease Control (CDC) *6:* 1180
Central Asia *1:* 198
Central Dogma *7:* 1283
Central Lowlands (North America) *7:* 1356
Centrifuge *3:* **445-446,** 446 (ill.)
Cephalopoda *7:* 1289
Cephalosporin *1:* 158
Cepheid variables *10:* 1964
Ceramic *3:* **447-448**
Cerebellum *2:* 345
Cerebral cortex *2:* 343
Cerebrum *2:* 343
Čerenkov effect *6:* 1189
Cerium *6:* 1163
Cesium *1:* 102
Cetaceans *3:* **448-451,** 450 (ill.), *4:* 681 (ill.), *7:* 1416 (ill.)
CFCs *6:* 1032, *7:* 1453-1454, *8:* 1555,
Chadwick, James *2:* 235, *7:* 1338
Chain, Ernst *1:* 157

Chamberlain, Owen *1:* 163
Chancroid *9:* 1735, 1736
Chandra X-ray Observatory *10:* 2040
Chandrasekhar, Subrahmanyan *10:* 1854
Chandrasekhar's limit *10:* 1854
Chao Phraya River *1:* 200
Chaos theory *3:* **451-453**
Chaparral *2:* 296
Chappe, Claude *10:* 1864
Chappe, Ignace *10:* 1864
Charles's law *5:* 961
Charon *8:* 1541, 1541 (ill.), 1542
Chassis *2:* 250
Cheetahs *5:* 861
Chemical bond *3:* **453-457**
Chemical compounds *3:* 541-546
Chemical elements *4:* 774-781
Chemical equations *5 :* 815-817
Chemical equilibrium *5:* 817-820
Chemical warfare *3:* **457-463,** 459 (ill.), 461 (ill.)*6:* 1032
Chemiluminescence *6:* 1198
Chemistry *3:* **463-469,** 465 (ill.) ,467 (ill.), *8:* 1603
Chemoreceptors *8:* 1484
Chemosynthesis *7:* 1418
Chemotherapy *2:* 382
Chichén Itzá *1:* 173
Chicxulub *1:* 202
Childbed fever *1:* 164
Chimpanzees *8:* 1572
Chiropractic *1:* 120
Chladni, Ernst *1:* 17
Chlamydia *9:* 1735, 1736
Chlorination *6:* 1033
Chlorine *6:* 1032
Chlorofluorocarbons. *See* **CFCs**
Chloroform *1:* 142, 143, 143 (ill.)
Chlorophyll *1:* 103
Chlorophyta *1:* 94
Chloroplasts *3:* 436, *8:* 1506 (ill.)
Chlorpromazine *10:* 1906
Cholesterol *3:* **469-471,** 471 (ill.), *6:* 1042
Chorionic villus sampling *2:* 322, *4:* 790
Chromatic aberration *10:* 1871
Chromatography *8:* 1604
Chromosomes *3:* **472-476,** 472 (ill.), 475 (ill.)

Index

Chromosphere *10:* 1846
Chrysalis *2:* 366, *7:* 1261 (ill.)
Chrysophyta *1:* 93
Chu, Paul Ching-Wu *10:* 1851
Cigarette smoke *3:* **476-478,** 477 (ill.)
Cigarettes, addiction to *1:* 34
Ciliophora *8:* 1592
Circle *3:* **478-480,** 479 (ill.)
Circular acceleration *1:* 5
Circular accelerators *8:* 1479
Circulatory system *3:* **480-484,** 482 (ill.)
Classical conditioning *9:* 1657
Clausius, Rudolf *10:* 1885
Claustrophobia *8:* 1497
Climax community *10:* 1839
Clones and cloning *3:* **484-490,** 486 (ill.), 489 (ill.)
Clostridium botulinum *2:* 258
Clostridium tetani *2:* 258
Clouds *3:* **490-492,** 491 (ill.)
Coal *3:* **492-498,** 496 (ill.)
Coast and beach *3:* **498-500.** 500 (ill.)
Coastal Plain (North America) *7:* 1356
Cobalt-60 *7:* 1373
COBE (Cosmic Background Explorer) *2:* 276
COBOL *3:* 551
Cocaine *1:* 34, *3:* **501-505,** 503 (ill.)
Cockroaches *3:* **505-508,** 507 (ill.)
Coelacanth *3:* **508-511,** 510 (ill.)
Cognition *3:* **511-515,** 513 (ill.), 514 (ill.)
Cold fronts *1:* 81, 81 (ill.)
Cold fusion *7:* 1371
Cold-deciduous forests *5:* 909
Collins, Francis *6:* 1064
Collins, Michael *9:* 1779
Colloids *3:* **515-517,** 517 (ill.)
Color *3:* **518-522,** 521 (ill.)
Color blindness *5:* 971
Colorant *4:* 686
Colt, Samuel *7:* 1237
Columbus, Christopher *1:* 63
Coma *2:* 345
Combined gas law *5:* 960
Combustion *3:* **522-527,** 524 (ill.), *7:* 1441
Comet Hale-Bopp *3:* 529
Comet Shoemaker-Levy 9 *6:* 1151
Comet, Halley's *3:* 528

Comets *3:* **527-531,** 529 (ill.), *6:* 1151, *9:* 1765
Common cold *10:* 1978
Compact disc *3:* **531-533,** 532 (ill.)
Comparative genomics *6:* 1067
Complex numbers *3:* **534-536,** 534 (ill.), *6:* 1082
Composite materials *3:* **536-539**
Composting *3:* **539-541,** 539 (ill.)
Compound, chemical *3:* **541-546,** 543 (ill.)
Compton Gamma Ray Observatory *5:* 949
Compulsion *7:* 1405
Computer Aided Design and Manufacturing. *See* **CAD/CAM**
Computer languages *1:* 189, *3:* 551
Computer software *3:* **549-554,** 553 (ill.)
Computer, analog *3:* **546-547**
Computer, digital *3:* **547-549,** 548 (ill.)
Computerized axial tomography. *See* **CAT scans**
Concave lenses *6:* 1185
Conditioning *9:* 1657
Condom *3:* 563
Conduction *6:* 1044
Conductivity, electrical. *See* **Electrical conductivity**
Conservation laws *3:* **554-558.** 557 (ill.)
Conservation of electric charge *3:* 556
Conservation of momentum *7:* 1290
Conservation of parity *3:* 558
Constellations *3:* **558-560,** 559 (ill.)
Contact lines *5:* 987
Continental Divide *7:* 1357
Continental drift *8:* 1534
Continental margin *3:* **560-562**
Continental rise *3:* 562
Continental shelf *2:* 300
Continental slope *3:* 561
Contraception *3:* **562-566,** 564 (ill.)
Convection *6:* 1044
Convention on International Trade in Endangered Species *5:* 795
Convex lenses *6:* 1185
Cooke, William Fothergill *10:* 1865
Coordination compounds *3:* 544
Copernican system *3:* 574
Copper *10:* 1919-1921, 1920 (ill.)

Index

Coral 3: 566-569, 567 (ill.), 568 (ill.)
Coral reefs 2: 301
Core 4: 711
Coriolis effect 2: 219, 10: 2029
Corona 10: 1846
Coronary artery disease 6: 1042
Coronas 2: 225
Correlation 3: 569-571
Corson, D. R. 6: 1035
Corti, Alfonso Giacomo Gaspare 4: 695
Corticosteroids 1: 206
Corundum 6: 1094
Cosmetic plastic surgery 8: 1530
Cosmic Background Explorer (COBE) 2: 276
Cosmic dust 6: 1130
Cosmic microwave background 2: 275, 8: 1637
Cosmic rays 3: 571-573, 573 (ill.)
Cosmology 1: 171, 3: 574-577
Cotton 3: 577-579, 578 (ill.)
Coulomb 3: 579-582
Coulomb, Charles 3: 579, 6: 1212
Coulomb's law 4: 744
Courtois, Bernard 6: 1035
Courtship behaviors 2: 273
Covalent bonding 3: 455
Cowan, Clyde 10: 1833
Coxwell, Henry Tracey 2: 263
Coyotes 2: 385
Craniotomy 8: 1528
Creationism 3: 577
Crick, Francis 3: 473, 4: 786, 5: 973, 980 (ill.), 982, 7: 1389
Cro-Magnon man 6: 1059
Crop rotation 3: 589
Crops 3: 582-590, 583 (ill.), 589 (ill.)
Crude oil 8: 1492
Crust 4: 709
Crustaceans 3: 590-593, 592 (ill.)
Cryobiology 3: 593-595
Cryogenics 3: 595-601, 597 (ill.)
Crystal 3: 601-604, 602 (ill.), 603 (ill.)
Curie, Marie 7: 1450
Current electricity 4: 742
Currents, ocean 3: 604-605
Cybernetics 3: 605-608, 607 (ill.)
Cyclamate 3: 608
Cyclone and anticyclone 3: 608-610, 609 (ill.)

Cyclotron 1: 163, 8: 1479, 1480 (ill.)
Cystic fibrosis 2: 320
Cytokinin 6: 1052
Cytoskeleton 3: 434

D

Da Vinci, Leonardo 2: 291, 4: 691, 10: 2020
Daddy longlegs 1: 171
Dalton, John 2: 226, 229, 2: 232
Dalton's theory 2: 232
Dam 4: 611-613, 612 (ill.)
Damselfly 1: 184 (ill.)
Danube River 5: 824
Dark matter 4: 613-616, 615 (ill.)
Dark nebulae 6: 1131, 7: 1330
Dart, Raymond 6: 1056
Darwin, Charles 1: 29, 6: 1051, 8: 1510
Dating techniques 4: 616-619, 618 (ill.)
Davy, Humphry 1: 142, chlorine 6: 1032, 1087
DDT (dichlorodiphenyltrichloroethane) 1: 69, 4: 619-622, 620 (ill.)
De Bort, Léon Philippe Teisserenc 2: 263
De Candolle, Augustin Pyrame 8: 1509
De curatorum chirurgia 8: 1528
De Forest, Lee 10: 1961
De materia medica 5: 877
De Soto, Hernando 7: 1299
Dead Sea 1: 196
Death 1: 59-62
Decay 7: 1442
Decimal system 1: 178
Decomposition 2: 392, 9: 1648
Deimos 6: 1229
Dementia 4: 622-624, 623 (ill.)
Democritus 2: 226, 232
Dendrochronology. *See* **Tree-ring dating**
Denitrification 7: 1343
Density 4: 624-626, 625 (ill.)
Dentistry 4: 626-630, 628 (ill.), 629 (ill.)
Depression 4: 630-634, 632 (ill.)
Depth perception 8: 1483 (ill.), 1484

Index

Dermis *6:* 1111
Desalination *7:* 1439, *10:* 2012
Descartes, René *6:* 1184
The Descent of Man *6:* 1055
Desert *2:* 296, *4:* **634-638,** 635 (ill.), 636 (ill.)
Detergents, soaps and *9:* 1756-1758
Devonian period *8:* 1461
Dew point *3:* 490
Dexedrine *2:* 238
Diabetes mellitus *4:* **638-640**
Diagnosis *4:* **640-644,** 643 (ill.)
Dialysis *4:* **644-646,** *7:* 1439
Diamond *2:* 396 (ill.), 397
Diencephalon *2:* 342
Diesel engine *4:* **646-647,** 647 (ill.)
Diesel, Rudolf *4:* 646, *10:* 1835
Differential calculus *2:* 372
Diffraction *4:* **648-651,** 648 (ill.)
Diffraction gratings *4:* 650
Diffusion *4:* **651-653,** 652 (ill.)
Digestion *7:* 1255
Digestive system *4:* **653-658,** 657 (ill.)
Digital audio tape *6:* 1211
Digital technology. *See* **Cellular/digital technology**
Dingoes *2:* 385, 385 (ill.)
Dinosaurs *4:* **658-665,** 660 (ill.), 663 (ill.), 664 (ill.)
Diodes *4:* **665-666,** *6:* 1176-1179
Dioscorides *5:* 878
Dioxin *4:* **667-669**
Dirac, Paul *1:* 163, *4:* 772
Dirac's hypothesis *1:* 163
Direct current (DC) *4:* 741
Dirigible *1:* 75
Disaccharides *2:* 388
Disassociation *7:* 1305
Disease *4:* **669-675,** 670 (ill.), 673 (ill.), *8:* 1518
Dissection *10:* 1989
Distillation *4:* **675-677,** 676 (ill.)
DNA *1:* 61, *2:* 310, *3:* 434, 473-474, *5:* 972-975, 980 (ill.), 981-984, *7:* 1389-1390
  forensic science *5:* 900
  human genome project *6:* 1060-1068
  mutation *7:* 1314-1316
Döbereiner, Johann Wolfgang *8:* 1486
Dogs. *See* **Canines**

Dollard, John *10:* 1871
Dolly (clone) *3:* 486
Dolphins *3:* 448, 449 (ill.)
Domagk, Gerhard *1:* 156
Domain names (computers) *6:* 1127
Dopamine *9:* 1720
Doppler effect *4:* **677-680,** 679 (ill.), *9:* 1654
Doppler radar *2:* 220 (ill.), *10:* 2023
Doppler, Christian Johann *9:* 1654
Down syndrome *2:* 319
Down, John Langdon Haydon *9:* 1713
Drake, Edwin L. *7:* 1419
Drebbel, Cornelius *10:* 1834
Drew, Richard *1:* 39
Drift nets *4:* **680-682,** 681 (ill.)
Drinker, Philip *8:* 1548
Drought *4:* **682-684,** 683 (ill.)
Dry cell (battery) *2:* 269
Dry ice *2:* 395
Drying (food preservation) *5:* 890
Dubois, Marie-Eugene *6:* 1058
Duodenum *4:* 655
Dura mater *2:* 342
Dust Bowl *4:* 682
Dust devils *10:* 1902
Dust mites *1:* 107, 108 (ill.)
DVD technology *4:* **684-686**
Dyes and pigments *4:* **686-690,** 688 (ill.)
Dynamite *5:* 845
Dysarthria *9:* 1798
Dyslexia *4:* **690-691,** 690 (ill.)
Dysphonia *9:* 1798
Dysprosium *6:* 1163

E

$E = mc^2$ *7:* 1363, 1366, *9:* 1662
Ear *4:* **693-698,** 696 (ill.)
Earth (planet) *4:* **698-702,** 699 (ill.)
Earth science *4:* **707-708**
Earth Summit *5:* 796
Earth's interior *4:* **708-711,** 710 (ill.)
Earthquake *4:* **702-707,** 705 (ill.), 706 (ill.)
Eating disorders *4:* **711-717,** 713 (ill.)
Ebola virus *4:* **717-720,** 719 (ill.)
Echolocation *4:* **720-722**
Eclipse *4:* **723-725,** 723 (ill.)

Index

Ecological pyramid *5:* 894 (ill.), 896
Ecological system. *See* **Ecosystem**
Ecologists *4:* 728
Ecology *4:* **725-728**
Ecosystem *4:* **728-730,** 729 (ill.)
Edison, Thomas Alva *6:* 1088
EEG (electroencephalogram) *2:* 348, *9:* 1746
Eijkman, Christian *10:* 1981
Einstein, Albert *4:* 691, *7:* 1428, *9:* 1659 (ill.)
  photoelectric effect *6:* 1188, *8:* 1504
  space-time continuum *9:* 1777
  theory of relativity *9:* 1659-1664
Einthoven, William *4:* 751
EKG (electrocardiogram) *4:* 751-755
El Niño *4:* **782-785,** 784 (ill.)
Elasticity *4:* **730-731**
Elbert, Mount *7:* 1357
Elbrus, Mount *5:* 823
Electric arc *4:* **734-737,** 735 (ill.)
Electric charge *4:* 743
Electric circuits *4:* 739, 740 (ill.)
Electric current *4:* 731, 734, **737-741,** 740 (ill.), 746, 748, 761, 767, 771, 773
Electric fields *4:* 743, 759
Electric motor *4:* **747-750,** 747 (ill.)
Electrical conductivity *4:* **731-734,** 735
Electrical force *3:* 579, 581-582, *4:* 744
Electrical resistance *4:* 732, 738, 746
Electricity *4:* **741-747,** 745 (ill.)
Electrocardiogram *4:* **751-755,** 753 (ill.), 754 (ill.)
Electrochemical cells *3:* 416, 436-439
Electrodialysis *4:* 646
Electroluminescence *6:* 1198
Electrolysis *4:* **755-758**
Electrolyte *4:* 755
Electrolytic cell *3:* 438
Electromagnet *6:* 1215
Electromagnetic field *4:* **758-760**
Electromagnetic induction *4:* **760-763,** 762 (ill.)
Electromagnetic radiation *8:* 1619
Electromagnetic spectrum *4:* **763-765,** *4:* 768, *6:* 1100, 1185, *8:* 1633, *9:* 1795
Electromagnetic waves *7:* 1268
Electromagnetism *4:* **766-768,** 766 (ill.)
Electron *4:* **768-773**
Electron gun *3:* 417
Electronegativity *3:* 455
Electronics *4:* **773-774,** 773 (ill.)
Electrons *4:* **768-773,** *10:* 1832, 1833
Electroplating *4:* 758
Element, chemical *4:* **774-781,** 778 (ill.)
Elementary algebra *1:* 98
Elements *4:* 775, 777, *8:* 1490, *10:* 1913
Embryo and embryonic development *4:* **785-791,** 788 (ill.)
Embryology *4:* 786
Embryonic transfer *4:* 790-791
Emphysema *9:* 1681
Encke division *9:* 1711
Encke, Johann *9:* 1711
Endangered species *5:* **793-796,** 795 (ill.)
Endangered Species Act *5:* 795
Endocrine system *5:* **796-801,** 799 (ill.)
Endoplasmic reticulum *3:* 433
Energy *5:* **801-805**
Energy and mass *9:* 1662
Energy conservation *1:* 117
Energy, alternative sources of *1:* 111-118, *6:* 1069
Engels, Friedrich *6:* 1097
Engineering *5:* **805-807,** 806 (ill.)
Engines *2:* 246, *6:* 1117, 1143, *9:* 1817, *10:* 1835
English units of measurement. *See* **British system of measurement**
ENIAC *3:* 551
Entropy *10:* 1886
Environment
  air pollution *8:* 1552, 1553
  effect of aerosols on *1:* 47, 48
  effect of carbon dioxide on *8:* 1554
  effect of use of fossil fuels on *2:* 285, *7:* 1454
  impact of aquaculture on *1:* 168
  industrial chemicals *8:* 1557
  ozone depletion *8:* 1555
  poisons and toxins *8:* 1546
  tropical deforestation *9:* 1744
  water pollution *8:* 1556
Environmental ethics *5:* **807-811,** 809 (ill.), 810 (ill.)

Index

Enzyme *5:* **812-815,** 812 (ill.), 814 (ill.)
Eosinophils *2:* 329
Epidemics *4:* 671
Epidermis *2:* 362, *6:* 1110
Epilepsy *2:* 347-349
Equation, chemical *5:* **815-817**
Equilibrium, chemical *5:* **817-820**
Equinox *9:* 1728
Erasistratus *1:* 138
Erbium *6:* 1163
Erosion *3:* 498, *5:* **820-823,** 821 (ill.), *9:* 1762
Erythroblastosis fetalis *9:* 1685
Erythrocytes *2:* 327
Escherichia coli *2:* 258
Esophagitis *4:* 656
Estrogen *5:* 801, *8:* 1599, 1600
Estuaries *2:* 300
Ethanol *1:* 89-91
Ether *1:* 142, 143
Ethics *3:* 489, *5:* 807-811
Ethylene glycol *1:* 91
Euglenoids *1:* 92
Euglenophyta *1:* 92
Eukaryotes *3:* 429, 432-435
Europa *6:* 1148, 1149
Europe *5:* **823-828,** 825 (ill.), 827 (ill.)
Europium *6:* 1163
Eutrophication *1:* 96, *5:* **828-831,** 830 (ill.)
Evans, Oliver *7:* 1237, *9:* 1820
Evaporation *5:* **831-832**
Everest, Mount *1:* 194
Evergreen broadleaf forests *5:* 909
Evergreen tropical rain forest *2:* 298
Evolution *1:* 26, 51, *5:* **832-839**
Excretory system *5:* **839-842**
Exhaust system *2:* 247
Exoplanets. *See* **Extrasolar planets**
Exosphere *2:* 214
Expansion, thermal *5:* 842-843, *10:* **1883-1884**
Expert systems *1:* 188
Explosives *5:* **843-847**
Extrasolar planets *5:* **847-848,** 846 (ill.)
Extreme Ultraviolet Explorer *6:* 1123
Exxon *Valdez* *7:* 1424, 1425 (ill.)
Eye *5:* **848-853,** 851 (ill.)
Eye surgery *8:* 1615-1618

F

Fahrenheit temperature scale *10:* 1882
Fahrenheit, Gabriel Daniel *10:* 1882
Far East *1:* 199
Faraday, Michael *4:* 761, 767, *6:* 1212
Farming. *See* **Agriculture**
Farnsworth, Philo *10:* 1875
Farsightedness *5:* 851
Father of
    acoustics *1:* 17
    American psychiatry *9:* 1713
    genetics *5:* 982
    heavier-than-air craft *1:* 77
    lunar topography *7:* 1296
    medicine *2:* 348
    modern chemistry *3:* 465
    modern dentistry *4:* 627
    modern evolutionary theory *5:* 833
    modern plastic surgery *8:* 1529
    rigid airships *1:* 75
    thermochemistry *3:* 525
Fats *6:* 1191
Fauchard, Pierre *4:* 627
Fault *5:* **855,** 856 (ill.)
Fault lines *5:* 987
Fear, abnormal or irrational. *See* **Phobias**
Feldspar *6:* 1094
Felines *5:* **855-864,** 861, 862 (ill.)
Fermat, Pierre de *7:* 1393, *8:* 1576
Fermat's last theorem *7:* 1393
Fermentation *5:* **864-867,** *10:* 2043
Fermi, Enrico *7:* 1365
Ferrell, William *2:* 218
Fertilization *5:* **867-870,** 868 (ill.)
Fertilizers *1:* 66
Fetal alcohol syndrome *1:* 87
Fiber optics *5:* **870-872,** 871 (ill.)
Fillings (dental) *4:* 628
Filovirus *4:* 717
Filtration *5:* **872-875**
Fingerprinting *5:* 900
Fire algae *1:* 94
First law of motion *6:* 1170
First law of planetary motion *7:* 1426
First law of thermodynamics *10:* 1885
Fish *5:* **875-878,** 876 (ill.)
Fish farming *1:* 166
Fishes, age of *8:* 1461
FitzGerald, George Francis *9:* 1660

Index

Flash lock *6:* 1193
Fleas *8:* 1474, 1474 (ill.)
Fleischmann, Martin *7:* 1371
Fleming, Alexander *1:* 156
Fleming, John Ambrose *10:* 1961
Florey, Howard *1:* 157
Flower *5: 878-862,* 881 (ill.)
Flu. *See* **Influenza**
Fluid dynamics *5: 882-886*
Flukes *8:* 1473
Fluorescence *6:* 1197
Fluorescent light *5: 886-888,* 888 (ill.)
Fluoridation *5: 889-890*
Fluoride *5:* 889
Fluorine *6:* 1031-1032
Fluorspar *6:* 1095
Fly shuttle *6:* 1097
Fold lines *5:* 987
Food irradiation *5:* 893
Food preservation *5: 890-894*
Food pyramid *7:* 1402, 1402 (ill.)
Food web and food chain *5: 894-898,* 896 (ill.)
Ford, Henry *2:* 249 (ill.), *7:* 1237-1238
Forensic science *5: 898-901,* 899 (ill.), *6:* 1067
Forestry *5: 901-907,* 905 (ill.), 906 (ill.)
Forests *2:* 294-295, *5: 907-914,* 909 (ill.), 910 (ill.), 913 (ill.)
Formula, chemical *5: 914-917*
FORTRAN *3:* 551
Fossil and fossilization *5: 917-921,* 919 (ill.), 920 (ill.), *6:* 1055, *7:* 1326 (ill.), *8:* 1458
Fossil fuels *1:* 112, *2:* 284, 392, *7:* 1319
Fossils, study of. *See* **Paleontology**
Foxes *2:* 384
Fractals *5: 921-923,* 922 (ill.)
Fractions, common *5: 923-924*
Fracture zones *7:* 1410
Francium *1:* 102
Free radicals *1:* 61
Freezing point *3:* 490
Frequency *4:* 763, *5: 925-926*
Frequency modulation *8:* 1628
Freshwater biomes *2:* 298
Freud, Sigmund *8:* 1593, 1594
Friction *5: 926-927*
Frisch, Otto *7:* 1362
Fronts *1:* 80-82
Fry, Arthur *1:* 39
Fujita Tornado Scale *10:* 1902
Fujita, T. Theodore *10:* 1902
Fuller, R. Buckminster *2:* 398
Fulton, Robert *10:* 1835
Functions (mathematics) *5: 927-930,* *8:* 1485
Functional groups *7:* 1430
Fungi *5: 930-934,* 932 (ill.)
Fungicides *1:* 67
Funk, Casimir *10:* 1982
Fyodorov, Svyatoslav N. *8:* 1617

G

Gabor, Dennis *6:* 1049
Gadolinium *6:* 1163
Gagarin, Yury *9:* 1778
Gaia hypothesis *5: 935-940*
Galactic clusters *9:* 1808
Galaxies, active *5:* 944
Galaxies *5: 941-945,* 941 (ill.), 943 (ill.), *9:* 1806-1808
Galen, Claudius *1:* 139
Galileo Galilei *1:* 4, *5:* 1012, *6: 1149, 1170, 1184,* *7: 1296,* *10: 1869*
Galileo probe *6:* 1149
Gall bladder *3:* 469, *4:* 653, 655
Galle, Johann *7:* 1330
Gallium *1:* 126
Gallo, Robert *10:* 1978
Gallstones *3:* 469
Galvani, Luigi *2:* 304, *4:* 751
Gambling *1:* 36
Game theory *5: 945-949*
Gamma rays *4:* 765, *5: 949-951,* *8:* 1632
Gamma-ray burst *5: 952-955,* 952 (ill.), 954 (ill.)
Ganges Plain *1:* 197
Ganymede *6:* 1148, 1149
Garbage. *See* **Waste management**
Gardening. *See* **Horticulture**
Gas, natural *7:* 1319-1321
Gases, electrical conductivity in *4:* 735
Gases, liquefaction of *5: 955-958*
Gases, properties of *5: 959-962,* 959 (ill.)
Gasohol *1:* 91

Gastropoda 7: 1288
Gauss, Carl Friedrich 6: 1212
Gay-Lussac, Joseph Louis 2: 262
Gay-Lussac's law 5: 962
Geiger counter 8: 1625
Gell-Mann, Murray 10: 1829
Generators 5: **962-966,** 964 (ill.)
Genes 7: 1248
Genes, mapping. See **Human Genome Project**
Genetic disorders 5: **966-973,** 968 (ill.), 968 (ill.)
Genetic engineering 2: 310, 5: **973-980,** 976 (ill.), 979 (ill.)
Genetic fingerprinting 5: 900
Genetics 5: **980-986,** 983 (ill.)
Geneva Protocol 2: 289
Genital herpes 9: 1735
Genital warts 9: 1735, 1737
Geocentric theory 3: 574
Geologic map 5: **986-989,** 988 (ill.)
Geologic time 5: **990-993**
Geologic time scale 5: 988
Geology 5: **993-994,** 944 (ill.)
Geometry 5: **995-999**
Geothermal energy 1: 116
Gerbert of Aurillac 7: 1396
Geriatrics 5: 999
Germ warfare. See **Biological warfare**
Germanium 2: 401
Gerontology 5: **999**
Gestalt psychology 8: 1595
Gibberellin 6: 1051
Gilbert, William 6: 1212
Gillies, Harold Delf 8: 1529
Gills 5: 877
Glacier 5: **1000-1003,** 1002 (ill.)
Glaisher, James 2: 263
Glass 5: **1004-1006,** 1004 (ill.)
Glenn, John 9: 1779
Gliders 1: 77
Global Biodiversity Strategy 2: 283
Global climate 5: **1006-1009**
Globular clusters 9: 1802, 1808
Glucose 2: 388
Gluons 10: 1831
Glutamate 9: 1720
Glycerol 1: 91
Glycogen 2: 389
Gobi Desert 1: 199
Goddard, Robert H. 9: 1695 (ill.)

Goiter 6: 1220
Gold 8: 1566-1569
Goldberger, Joseph 6: 1219
Golden-brown algae 1: 93
Golgi body 3: 433
Gondwanaland 1: 149
Gonorrhea 9: 1735, 1736
Gorillas 8: 1572
Gould, Stephen Jay 1: 32
Graphs and graphing 5: **1009-1011**
Grasslands 2: 296
Gravitons 10: 1831
Gravity and gravitation 5: **1012-1016,** 1014 (ill.)
Gray, Elisha 10: 1867
Great Barrier Reef 2: 240
Great Dividing Range 2: 240
Great Lakes 6: 1159
Great Plains 4: 682, 7: 1356
Great Red Spot (Jupiter) 6: 1149, 1150 (ill.)
Great Rift Valley 1: 49, 51
Great White Spot (Saturn) 9: 1709
Green algae 1: 94
Green flashes 2: 224
Greenhouse effect 2: 285, 393, 5: 1003, **1016-1022,** 1020 (ill.). 8: 1554, 10: 1965
Gregorian calendar 2: 373, 375
Grissom, Virgil 9: 1779
Growth hormone 5: 797
Growth rings (trees) 4: 619
Guiana Highlands 9: 1772
Guided imagery 1: 119
Gum disease 4: 630
Guth, Alan 2: 276
Gymnophions 1: 137
Gymnosperms 9: 1729
Gynecology 5: **1022-1024,** 1022 (ill.)
Gyroscope 5: **1024-1025,** 1024 (ill.)

H

H.M.S. *Challenger* 7: *1413*
Haber process 7: 1346
Haber, Fritz 7: 1346
Hadley, George 2: 218
Hahn, Otto 7: 1361
Hale, Alan 3: 529
Hale-Bopp comet 3: 529

Index

Hales, Stephen *2:* 337
Half-life *6:* **1027**
Halite *6:* 1096
Hall, Charles M. *1:* 124, *4:* 757
Hall, Chester Moore *10:* 1871
Hall, John *7:* 1237
Halley, Edmond *7:* 1262, *10:* 2020
Halley's comet *3:* 528
Hallucinogens *6:* **1027-1030**
Haloes *2:* 224
Halogens *6:* **1030-1036**
Hand tools *6:* **1036-1037,** 1036 (ill.)
Hard water *9:* 1757
Hargreaves, James *6:* 1098
Harmonices mundi *3:* 425
Harmonics *5:* 925
Hart, William Aaron *7:* 1320
Harvestmen (spider) *1:* 171, 170 (ill.)
Harvey, William *1:* 139, *2:* 292
Hazardous waste *10:* 2006-2007, 2006 (ill.)
HDTV *10:* 1879
Heart *6:* **1037-1043,** 1041 (ill.), 1042 (ill.)
Heart attack *6:* 1043
Heart diseases *3:* 470, *6:* 1040
Heart transplants *10:* 1926
Heart, measure of electrical activity. *See* **Electrocardiogram**
Heartburn *4:* 656
Heat *6:* **1043-1046**
Heat transfer *6:* 1044
Heat, measurement of. *See* **Calorie**
Heisenberg, Werner *8:* 1609
Heliocentric theory *3:* 574
Helium *7:* 1349
Helminths *8:* 1471 (ill.)
Hemiptera *6:* 1105
Hemodialysis *5:* 841
Henbury Craters *2:* 240
Henry, Joseph *4:* 761, *10:* 1865
Herbal medicine *1:* 120
Herbicides *1:* 54-59
Herculaneum *5:* 828
Heredity *7:* 1246
Hermaphroditism *9:* 1667
Heroin *1:* 32, 34
Herophilus *1:* 138
Héroult, Paul *1:* 124
Herpes *9:* 1737
Herschel, John *2:* 277

Herschel, William *2:* 277, *10:* 1871, 1952
Hertz, Heinrich *6:* 1188, *8:* 1502, 1626
Hess, Henri *3:* 525
Hevelius, Johannes *7:* 1296
Hewish, Antony *7:* 1340
Hibernation *6:* **1046-1048,** 1047 (ill.)
Himalayan Mountains *1:* 194, 197, *7:* 1301
Hindbrain *2:* 340
Hindenburg *1:* 76
Hindu-Arabic number system *1:* 178, *7:* 1396, *10:* 2047
Hippocrates *2:* 348
Histamine *6:* 1085
Histology *1:* 141
Historical concepts *6:* 1186
HIV (human immunodeficiency virus) *1:* 70, 72 (ill.), *8:* 1583
Hodgkin's lymphoma *6:* 1201
Hoffmann, Felix *1:* 6
Hofstadter, Robert *7:* 1339
Hogg, Helen Sawyer *10:* 1964
Holistic medicine *1:* 120
Holland, John *10:* 1835
Hollerith, Herman *3:* 549
Holmium *6:* 1163
Holograms and holography *6:* **1048-1050,** 1049 (ill.)
Homeopathy *1:* 120
Homeostasis *8:* 1516, 1517
Homo erectus *6:* 1058
Homo ergaster *6:* 1058
Homo habilis *6:* 1058
Homo sapiens *6:* 1055, 1058-1059
Homo sapiens sapiens *6:* 1059
Hooke, Robert *1:* 140, *4:* 731
Hooke's law *4:* 731
Hopewell mounds *7:* 1301
Hopkins, Frederick G. *10:* 1982
Hopper, Grace *3:* 551
Hormones *6:* **1050-1053**
Horticulture *6:* **1053-1054,** 1053 (ill.)
HTTP *6:* 1128
Hubble Space Telescope *9:* 1808, *10:* 1873, 1872 (ill.)
Hubble, Edwin *2:* 275, *7:* 1328, *9:* 1655, 1810
Human evolution *6:* **1054-1060,** 1057 (ill.), 1059 (ill.)
Human Genome Project *6:* **1060-**

Index

1068, 1062 (ill.), 1065 (ill.), 1066 (ill.)
Human-dominated biomes   *2:* 302
Humanistic psychology   *8:* 1596
Humason, Milton   *9:* 1655
Hurricanes   *3:* 610
Hutton, James   *10:* 1947
Huygens, Christiaan   *6:* 1187, *9:* 1711
Hybridization   *2:* 310
Hydrocarbons   *7:* 1430-1431
Hydrogen   *6:* **1068-1071,** 1068 (ill.)
Hydrologic cycle   *6:* **1071-1075,** 1072 (ill.), 1073 (ill.)
Hydropower   *1:* 113
Hydrosphere   *2:* 305
Hydrothermal vents   *7:* 1418, 1417 (ill.)
Hygrometer   *10:* 2020
Hypertension   *3:* 484
Hypertext   *6:* 1128
Hypnotherapy   *1:* 119
Hypotenuse   *10:* 1932
Hypothalamus   *2:* 342, 343
Hypothesis   *9:* 1723

I

Icarus   *1:* 74
Ice ages   *6:* **1075-1078,** 1077 (ill.)
Icebergs   *6:* **1078-1081,** 1080 (ill.), 1081 (ill.)
Idiot savants.   *See* **Savants**
IgE   *1:* 109
Igneous rock   *9:* 1702
Ileum   *4:* 656
Imaginary numbers   *6:* **1081-1082**
Immune system   *1:* 108, *6:* **1082-1087**
Immunization   *1:* 161, *10:* 1060-1960
Immunoglobulins   *1:* 159
Imprinting   *2:* 272
In vitro fertilization   *4:* 791
Incandescent light   *6:* **1087-1090,** 1089 (ill.)
Inclined plane   *6:* 1207
Indian peninsula   *1:* 197
Indicator species   *6:* **1090-1092,** 1091 (ill.)
Indium   *1:* 126
Induction   *4:* 760

Industrial minerals   *6:* 1092-1097
Industrial Revolution   *1:* 28, *3:* 523, *6:* 1193, **1097-1100,** *7:* 1236, *9:* 1817
  automation   *2:* 242
  effect on agriculture   *1:* 63
  food preservation   *5:* 892
Infantile paralysis.   *See* **Poliomyelitis**
Infants, sudden death.   *See* **Sudden infant death syndrome (SIDS)**
Inflationary theory   *2:* 275, 276
Influenza   *4:* 672, *6:* 1084, *10:* 1978, 1979-1981
Infrared Astronomical Satellite   *9:* 1808
Infrared astronomy   *6:* **1100-1103,** 1102 (ill.)
Infrared telescopes   *6:* 1101
Ingestion   *4:* 653
Inheritance, laws of.   *See* **Mendelian laws of inheritance**
Insecticides   *1:* 67
Insects   *6:* **1103-1106,** 1104 (ill.)
Insomnia   *9:* 1747
Insulin   *3:* 474, *4:* 638
Integers   *1:* 180
Integral calculus   *2:* 372
Integrated circuits   *6:* **1106-1109,** 1108 (ill.), 1109 (ill.)
Integumentary system   *6:* **1109-1112,** 1111 (ill.)
Interference   *6:* **1112-1114,** 1113 (ill.)
Interferometer   *6:* 1115 (ill.), 1116
Interferometry   *10:* 1874, *6:* **1114-1116,** 1115 (ill.), 1116 (ill.)
Interferon   *6:* 1084
Internal-combustion engines   *6:* **1117-1119,** 1118 (ill.)
International Space Station   *9:* 1788
International System of Units   *2:* 376
International Ultraviolet Explorer   *10:* 1946, *6:* **1120-1123,** 1122 (ill.)
Internet   *6:* **1123-1130,** 1127 (ill.)
Interstellar matter   *6:* **1130-1133,** 1132 (ill.)
Invertebrates   *6:* **1133-1134,** 1134 (ill.)
Invertebrates, age of   *8:* 1461
Io   *6:* 1148, 1149
Iodine   *6:* 1035
Ionic bonding   *3:* 455
Ionization   *6:* **1135-1137**
Ionization energy   *6:* 1135

U·X·L Encyclopedia of Science, 2nd Edition   xlix

Index

Ions *4:* 733
Iron *10:* 1915-1918
Iron lung *8:* 1548 (ill.)
Iron manufacture *6:* 1098
Irrational numbers *1:* 180, 181
Isaacs, Alick *6:* 1084
Islands *3:* 500, *6:* **1137-1141**, 1139 (ill.)
Isotopes *6:* **1141-1142**, *7:* 1241
IUE. *See* **International Ultraviolet Explorer**

J

Jackals *2:* 385
Jacquet-Droz, Henri *9:* 1691
Jacquet-Droz, Pierre *9:* 1691
James, William *8:* 1594
Jansky, Karl *8:* 1635
Java man *6:* 1058
Jefferson, Thomas *7:* 1300
Jejunum *4:* 655
Jenner, Edward *1:* 161, *10:* 1957
Jet engines *6:* **1143-1146**, 1143 (ill.), 1145 (ill.)
Jet streams *2:* 221, *4:* 783, *7:* 1293
Jones, John *8:* 1529
Joule *6:* 1045
Joule, James *10:* 1885
Joule-Thomson effect *3:* 597
Jupiter (planet) *6:* **1146-1151**, 1147 (ill.), 1150 (ill.)

K

Kangaroos and wallabies *6:* **1153-1157**, 1155 (ill.)
Kant, Immanuel *9:* 1765
Kay, John *6:* 1097
Kelvin scale *10:* 1882
Kelvin, Lord. *See* **Thomson, William**
Kenyanthropus platyops *6:* 1056
Kepler, Johannes *3:* 425, 574, *7:* 1426
Keratin *6:* 1110
Kettlewell, Henry Bernard David *1:* 28
Kidney dialysis *4:* 645
Kidney stones *5:* 841
Kilimanjaro, Mount *5:* 1000
Kinetic theory of matter *7:* 1243

King, Charles G. *6:* 1219
Klein bottle *10:* 1899
Knowing. *See* **Cognition**
Klein, Felix *10:* 1899
Koch, Robert *2:* 292
Köhler, Wolfgang *8:* 1595
Kraepelin, Emil *9:* 1718
Krakatoa *10:* 1998
Krypton *7:* 1349, 1352
Kuiper Disk *3:* 530
Kuiper, Gerald *3:* 530, *7:* 1333
Kwashiorkor *6:* 1218, *7:* 1403

L

La Niña *4:* 782
Lacrimal gland *5:* 852
Lactose *2:* 388
Lager *2:* 354
Lake Baikal *1:* 198
Lake Huron *6:* 1162 (ill.)
Lake Ladoga *5:* 824
Lake Michigan *7:* 1354 (ill.)
Lake Superior *6:* 1159
Lake Titicaca *6:* 1159
Lakes *6:* **1159-1163**, 1161 (ill.), 1162 (ill.)
Lamarck, Jean-Baptiste *1:* 28
Lambert Glacier *1:* 149
Laminar flow *1:* 40
Landfills *10:* 2007, 2008 (ill.)
Language *3:* 515
Lanthanides *6:* **1163-1166**
Lanthanum *6:* 1163
Laplace, Pierre-Simon *2:* 323, *9:* 1765
Large intestine *4:* 656
Laryngitis *9:* 1681
Laser eye surgery *8:* 1617
Lasers *6:* **1166-1168**, 1168 (ill.)
LASIK surgery *8:* 1617
Laurentian Plateau *7:* 1355
Lava *10:* 1995
Lavoisier, Antoine Laurent *3:* 465, 524, *6:* 1069, *7:* 1444
Law of conservation of energy *3:* 555, *10:* 1885
Law of conservation of mass/matter *3:* 554, *5:* 816
Law of conservation of momentum *7:* 1290

Index

Law of dominance *7:* 1249
Law of electrical force *3:* 579
Law of independent assortment *7:* 1249
Law of planetary motion *3:* 425
Law of segregation *7:* 1249
Law of universal gravitation *3:* 426, *7:* 1427
Lawrence, Ernest Orlando *8:* 1479
Laws of motion *3:* 426, *6:* **1169-1171,** *7:* 1235, 1426
Le Verrier, Urbain *7:* 1330
Lead *2:* 402-403
Leakey, Louis S. B. *6:* 1058
Learning disorders *4:* 690
Leaf *6:* **1172-1176,** 1172 (ill.), 1174 (ill.)
Leavitt, Henrietta Swan *10:* 1964
Leclanché, Georges *2:* 269
LED (light-emitting diode) *6:* **1176-1179,** 1177 (ill.), 1178 (ill.)
Leeuwenhoek, Anton van *2:* 253, *6:* 1184, *8:* 1469
Legionella pneumophilia *6:* 1181, 1182
Legionnaire's disease *6:* **1179-1184,** 1182 (ill.)
Leibniz, Gottfried Wilhelm *2:* 371, 372, *7:* 1242
Lemaître, Georges-Henri *3:* 576
Lemurs *8:* 1572
Lenoir, Jean-Joseph Éttien *6:* 1119
Lenses *6:* **1184-1185,** 1184 (ill.)
Lentic biome *2:* 298
Leonid meteors *7:* 1263
Leonov, Alexei *9:* 1779
Leopards *5:* 860
Leptons *10:* 1830
Leucippus *2:* 226
Leukemia *2:* 380
Leukocytes *2:* 328
Lever *6:* 1205, 1206 (ill.)
Levy, David *6:* 1151
Lewis, Gilbert Newton *1:* 15
Liber de Ludo Aleae *8:* 1576
Lice *8:* 1474
Life, origin of *4:* 702
Light *6:* 1087-1090, **1185-1190**
Light, speed of *6:* 1190
Light-year *6:* **1190-1191**
Lightning *10:* 1889, 1889 (ill.)
Limbic system *2:* 345
Liming agents *1:* 66

Lind, James *6:* 1218, *10:* 1981
Lindenmann, Jean *6:* 1084
Linear acceleration *1:* 4
Linear accelerators *8:* 1477
Linnaeus, Carolus *2:* 292, 337
Lions *5:* 860
Lipids *6:* **1191-1192,** *7:* 1400
Lippershey, Hans *10:* 1869
Liquid crystals *7:* 1244-1245
Liquor. *See* **Alcohol (liquor)**
Lister, Joseph *1:* 165
Lithium *1:* 100
Lithosphere *8:* 1535, 1536
Litmus test *8:* 1496
Locks (water) *6:* **1192-1195,** 1193 (ill.)
Logarithms *6:* **1195**
Long, Crawford W. *1:* 143
Longisquama insignis *2:* 312
Longitudinal wave *10:* 2015
Lord Kelvin. *See* **Thomson, William**
Lorises *8:* 1572
Lotic *2:* 299
Lotic biome *2:* 299
Lowell, Percival *8:* 1539
Lowry, Thomas *1:* 15
LSD *6:* 1029
Lucy (fossil) *6:* 1056, 1057 (ill.)
Luminescence *6:* **1196-1198,** 1196 (ill.)
Luna *7:* 1296
Lunar eclipses *4:* 725
Lunar Prospector *7:* 1297
Lung cancer *9:* 1682
Lungs *9:* 1679
Lunisolar calendar *2:* 374
Lutetium *6:* 1163
Lymph *6:* 1199
Lymph nodes *6:* 1200
Lymphatic system *6:* **1198-1202**
Lymphocytes *2:* 329, *6:* 1085, 1200 (ill.)
Lymphoma *2:* 380, *6:* 1201
Lysergic acid diethylamide. *See* **LSD**

M

Mach number *5:* 883
Mach, L. *6:* 1116
Machines, simple *6:* **1203-1209,** 1206 (ill.), 1208 (ill.)

Index

Mackenzie, K. R. *6:* 1035
Magellan *10:* 1966
Magma *10:* 1995
Magnesium *1:* 103
Magnetic fields *4:* 759
Magnetic fields, stellar. See **Stellar magnetic fields**
Magnetic recording/audiocassette *6:* **1209-1212,** 1209 (ill.), 1211 (ill.)
Magnetic resonance imaging *2:* 304
Magnetism *6:* **1212-1215,** 1214 (ill.)
Malnutrition *6:* **1216-1222,** 1221 (ill.)
Malpighi, Marcello *1:* 139
Mammals *6:* **1222-1224,** 1223 (ill.)
Mammals, age of *8:* 1462
Mangrove forests *5:* 909
Manhattan Project *7:* 1365, 1380
Manic-depressive illness *4:* 631
Mantle *4:* 710
Manufacturing. See **Mass production**
MAP (Microwave Anisotroy Probe) *2:* 276
Maps and mapmaking. See **Cartography**
Marasmus *6:* 1218
Marconi, Guglielmo *8:* 1626
Marie-Davy, Edme Hippolyte *10:* 2021
Marijuana *6:* **1224-1227,** 1226 (ill.), 1227 (ill.)
Marine biomes *2:* 299
Mariner 10 *7:* 1250
Mars (planet) *6:* **1228-1234,** 1228 (ill.), 1231 (ill.), 1232 (ill.)
Mars Global Surveyor *6:* 1230
Mars Pathfinder *6:* 1232
Marshes *10:* 2025
Maslow, Abraham *8:* 1596
Mass *7:* **1235-1236**
Mass production *7:* **1236-1239,** 1238 (ill.)
Mass spectrometry *7:* **1239-1241,** 1240 (ill.), *8:* 1604
Mastigophora *8:* 1592
Mathematics *7:* **1241-1242**
  imaginary numbers *6:* 1081
  logarithms *6:* 1195
  multiplication *7:* 1307
  number theory *7:* 1393
  probability theory *8:* 1575
  proofs *8:* 1578
  statistics *9:* 1810
  symbolic logic *10:* 1859
  topology *10:* 1897-1899
  trigonometry *10:* 1931-1933
  zero *10:* 2047
Matter, states of *7:* **1243-1246,** 1243 (ill.)
Maxwell, James Clerk *4:* 760, 767, *6:* 1213, *8:* 1626
Maxwell's equations *4:* 760
McKay, Frederick *5:* 889
McKinley, Mount *7:* 1302
McMillan, Edwin *1:* 24
Measurement. See **Units and standards**
Mechanoreceptors *8:* 1484
Meditation *1:* 119
Medulla oblongata *2:* 340
Meiosis *9:* 1666
Meissner effect *10:* 1851 (ill.)
Meitner, Lise *7:* 1362
Mekong River *1:* 200
Melanin *6:* 1110
Melanomas *2:* 380
Memory *2:* 344, *3:* 515
Mendel, Gregor *2:* 337, *4:* 786, *7:* 1247
Mendeleev, Dmitry *4:* 777, *8:* 1487
Mendelian laws of inheritance *7:* **1246-1250,** 1248 (ill.)
Meninges *2:* 342
Menopause *1:* 59, *2:* 410, *5:* 800, 1020
Menstruation *1:* 59, *5:* 800, 1020, *8:* 1599
Mental illness, study and treatment of. See **Psychiatry**
Mercalli scale *4:* 704
Mercalli, Guiseppe *4:* 704
Mercury *10:* 1921-1923, 1922 (ill.)
Mercury (planet) *7:* **1250-1255,** 1251 (ill.), 1252 (ill.)
Mercury barometers *2:* 265
Méré, Chevalier de *8:* 1576
Mescaline *6:* 1029
Mesosphere *2:* 213
Mesozoic era *5:* 990, *8:* 1462
Metabolic disorders *7:* **1254-1255,** 1254 (ill.), 1257
Metabolism *7:* **1255-1259**
Metalloids *7:* 1348
Metamorphic rocks *9:* 1705
Metamorphosis *7:* **1259-1261**

Index

Meteorograph *2:* 215
Meteors and meteorites *7:* **1262-1264**
Methanol *1:* 89
Metric system *7:* **1265-1268,** *10:* 1949
Mettauer, John Peter *8:* 1529
Meyer, Julius Lothar *4:* 777, *8:* 1487
Michell, John *2:* 323
Michelson, Albert A. *6:* 1114, *6:* 1187
Microwave Anisotropy Probe (MAP) *2:* 276
Microwave communication *7:* **1268-1271,** 1270 (ill.)
Microwaves *4:* 765
Mid-Atlantic Ridge *7:* 1303, 1409
Midbrain *2:* 340
Middle East *1:* 196
Mifepristone *3:* 565
Migraine *2:* 349, 350
Migration (animals) *7:* **1271-1273,** 1272 (ill.)
Millennium *2:* 375
Millikan, Robert Andrew *4:* 771
Minerals *6:* 1092-1097, *7:* 1401, **1273-1278,** 1276 (ill.), 1277 (ill.)
Mining *7:* **1278-1282,** 1281 (ill.)
Mir *9:* 1781
Mirages *2:* 222
Miranda *10:* 1954
Misch metal *6:* 1165
Missiles. *See* **Rockets and missiles**
Mission *9:* 1787
Mississippi River *7:* 1355
Mississippian earthern mounds *7:* 1301
Missouri River *7:* 1355
Mitchell, Mount *7:* 1356
Mites *1:* 170
Mitochondira *3:* 436
Mitosis *1:* 133, *9:* 1665
Mobile telephones *3:* 441
Möbius strip *10:* 1899
Möbius, Augustus Ferdinand *10:* 1899
Model T (automobile) *7:* 1237, 1238
Modulation *8:* 1627
Moho *4:* 709
Mohorovičiá discontinuity *4:* 709
Mohorovičiá, Andrija *4:* 709
Mohs scale *1:* 3
Mohs, Friedrich *1:* 3
Mole (measurement) *7:* **1282-1283**
Molecular biology *7:* **1283-1285**
Molecules *7:* **1285-1288,** 1286 (ill.)

Mollusks *7:* **1288-1290,** 1289 (ill.)
Momentum *7:* **1290-1291**
Monkeys *8:* 1572
Monoclonal antibodies *1:* 162, *2:* 311
Monocytes *2:* 329
Monoplacophora *7:* 1289
Monosaccharides *2:* 388
Monotremes *6:* 1224
Monsoons *7:* **1291-1294**
Mont Blanc *5:* 827
Montgolfier, Jacques *2:* 262
Montgolfier, Joseph *2:* 262
Moon *7:* **1294-1298,** 1295 (ill.), 1297 (ill.)
  affect on tides *10:* 1890
  Apollo 11 *9:* 1779
Morley, Edward D. *6:* 1187
Morphine *1:* 32, 33
Morse code *10:* 1866
Morse, Samuel F. B. *10:* 1865
Morton, William *1:* 143
Mosquitoes *6:* 1106, *8:* 1473
Motion, planetary, laws of *7:* 1426
Motors, electric. *See* **Electric motors**
Mounds, earthen *7:* **1298-1301,** 1299 (ill.)
Mount Ararat *1:* 197
Mount Cameroon *1:* 51
Mount Elbert *7:* 1357
Mount Elbrus *5:* 823
Mount Everest *1:* 194
Mount Kilimanjaro *1:* 53 (ill.), *5:* 1003, *7:* 1303
Mount McKinley *7:* 1302, 1354
Mount Mitchell *7:* 1356
Mount Robson *7:* 1357
Mount St. Helens *10:* 1996, 1998 (ill.)
Mountains *7:* **1301-1305,** 1304 (ill.)
Movable bridges *2:* 358
mRNA *7:* 1285
Müller, Karl Alex *10:* 1851
Multiple personality disorder *7:* **1305-1307**
Multiplication *7:* **1307-1309**
Muscular dystrophy *7:* 1313, 1337
Muscular system *7:* **1309-1313,** 1311 (ill.), 1312 (ill.), 1313 (ill.)
Mushrooms *6:* 1028
Mutation *7:* **1314-1317,** 1316 (ill.)
Mysophobia *8:* 1497

Index

N

Napier, John *6:* 1195
Narcolepsy *9:* 1748
Narcotics *1:* 32
Natural gas *7:* **1319-1321**
Natural language processing *1:* 189
Natural numbers *1:* 180, *7:* **1321-1322**
Natural selection *1:* 29-30, *2:* 292, *5:* 834, 837-839
Natural theology *1:* 27
Naturopathy *1:* 122
Nautical archaeology *7:* **1323-1327**, 1325 (ill.), 1326 (ill.)
Nautilus *10:* 1835
Navigation (animals) *7:* 1273
Neanderthal man *6:* 1059
Neap tides *10:* 1892
NEAR Shoemaker *1:* 203-204, *9:* 1787
Nearsightedness *5:* 851
Nebula *7:* **1327-1330**, 1329 (ill.)
Nebular hypothesis *9:* 1765
Negative reinforcement. *See* **Reinforcement, positive and negative**
Nematodes *8:* 1471
Neodymium *6:* 1163
Neolithic Age *1:* 62
Neon *7:* 1349, 1352
Neptune (planet) *7:* **1330-1333**, 1331 (ill.)
Nereid *7:* 1333
Nerve nets *7:* 1333
Nervous system *7:* **1333-1337**, 1335 (ill.)
Neurons *7:* 1333 (ill.)
Neurotransmitters *1:* 128, *2:* 350, *4:* 631, *7:* 1311, *9:* 1720
Neutralization *1:* 15
Neutrinos *10:* 1832, 1833
Neutron *2:* 235, *7:* **1337-1339**, *10:* 1830, 1832
Neutron stars *7:* **1339-1341**, 1341 (ill.)
Neutrophils *2:* 329
Newcomen, Thomas *9:* 1818
Newton, Isaac *1:* 4, *5:* 1012, *6:* 1184, *7:* 1242
    calculus *2:* 372
    corpsucular theory of light *6:* 1187
    laws of motion *6:* 1169-1171, *7:* 1427

reflector telescope *10:* 1871
Ngorongoro Crater *1:* 52
Niacin. *See* **Vitamin B3**
Nicotine *1:* 34, *3:* 478
Night blindness *6:* 1220
Nile River *9:* 1686
Nipkow, Paul *10:* 1875
Nitrification *7:* 1343, 1344
Nitrogen *7:* 1344, 1345 (ill.)
Nitrogen cycle *7:* **1342-1344**, 1342 (ill.)
Nitrogen family *7:* **1344-1349**, 1345 (ill.)
Nitrogen fixation *7:* 1342
Nitroglycerin *5:* 844
Nitrous oxide *1:* 142
Nobel, Alfred *5:* 845
Noble gases *7:* **1349-1352**, 1350 (ill.)
Non-Hodgkin's lymphoma *6:* 1201
Nondestructive testing *10:* 2037
North America *7:* **1352-1358**, 1353 (ill.), 1354 (ill.), 1357 (ill.)
Novas *7:* **1359-1360**, 1359 (ill.)
NSFNET *6:* 1127
Nuclear fission *7:* **1361-1366**, 1365 (ill.), 1381
Nuclear fusion *7:* **1366-1371**, 1370 (ill.)
Nuclear medicine *7:* **1372-1374**
Nuclear Non-Proliferation Treaty *7:* 1387
Nuclear power *1:* 113, *7:* **1374-1381**, 1376 (ill.), 1379 (ill.), *10:* 1836
Nuclear power plant *7:* 1374, 1379 (ill.)
Nuclear reactor *7:* 1365
Nuclear waste management *7:* 1379
Nuclear weapons *7:* **1381-1387**
Nucleic acids *7:* **1387-1392**, 1391 (ill.), 1392 (ill.)
Nucleotides *3:* 473, *7:* 1388
Nucleus (cell) *3:* 434
Number theory *7:* 1322, **1393-1395**
Numbers, imaginary *6:* 1081
Numbers, natural *7:* 1321
Numeration systems *7:* **1395-1399**
Nutrient cycle *2:* 307
Nutrients *7:* 1399
Nutrition *6:* 1216, *7:* **1399-1403**, 1402 (ill.)
Nylon *1:* 186, *8:* 1533

O

Oberon *10:* 1954
Obesity *4:* 716
Obsession *7:* **1405-1407**
Obsessive-compulsive disorder *7:* 1405
Obsessive-compulsive personality disorder *7:* 1406
Occluded fronts *1:* 82
Ocean *7:* **1407-1411,** 1407 (ill.)
Ocean currents *3:* 604-605
Ocean ridges *7:* 1410 (ill.)
Ocean zones *7:* **1414-1418**
Oceanic archaeology. *See* **Nautical archaeology**
Oceanic ridges *7:* 1409
Oceanography *7:* **1411-1414,** 1412 (ill.), 1413 (ill.)
Octopus *7:* 1289
Oersted, Hans Christian *1:* 124, *4:* 760, 766, *6:* 1212
Offshore drilling *7:* 1421
Ohio River *7:* 1355
Ohm (Ω) *4:* 738
Ohm, Georg Simon *4:* 738
Ohm's law *4:* 740
Oil drilling *7:* **1418-1422,** 1420 (ill.)
Oil pollution *7:* 1424
Oil spills *7:* **1422-1426,** 1422 (ill.), 1425 (ill.)
Oils *6:* 1191
Olduvai Gorge *6:* 1058
Olfaction. *See* **Smell**
On the Origin of Species by Means of Natural Selection *6:* 1054
On the Structure of the Human Body *1:* 139
O'Neill, J. A. *6:* 1211
Onnes, Heike Kamerlingh *10:* 1850
Oort cloud *3:* 530
Oort, Jan *8:* 1637
Open clusters *9:* 1808
Open ocean biome *2:* 299
Operant conditioning *9:* 1658
Ophediophobia *8:* 1497
Opiates *1:* 32
Opium *1:* 32, 33
Orangutans *8:* 1572, 1574 (ill.)
Orbit *7:* **1426-1428**
Organ of Corti *4:* 695
Organic chemistry *7:* **1428-1431**
Organic families *7:* 1430
Organic farming *7:* **1431-1434,** 1433 (ill.)
Origin of life *4:* 702
Origins of algebra *1:* 97
Orizaba, Pico de *7:* 1359
Orthopedics *7:* **1434-1436**
Oscilloscopes *10:* 1962
Osmosis *4:* 652, *7:* **1436-1439,** 1437 (ill.)
Osmotic pressure *7:* 1436
Osteoarthritis *1:* 181
Osteoporosis *9:* 1742
Otitis media *4:* 697
Otosclerosis *4:* 697
Ovaries *5:* 800
Oxbow lakes *6:* 1160
Oxidation-reduction reactions *7:* **1439-1442,** *9:* 1648
Oxone layer *7:* 1452 (ill.)
Oxygen family *7:* **1442-1450,** 1448 (ill.)
Ozone *7:* **1450-1455,** 1452 (ill.)
Ozone depletion *1:* 48, *8:* 1555
Ozone layer *7:* 1451

P

Packet switching *6:* 1124
Pain *7:* 1336
Paleoecology *8:* **1457-1459,** 1458 (ill.)
Paleontology *8:* **1459-1462,** 1461 (ill.)
Paleozoic era *5:* 990, *8:* 1461
Paleozoology *8:* 1459
Panama Canal *6:* 1194
Pancreas *4:* 655, *5:* 798
Pangaea *8:* 1534, 1536 (ill.)
Pap test *5:* 1020
Papanicolaou, George *5:* 1020
Paper *8:* **1462-1467,** 1464 (ill.), 1465 (ill.), 1466 (ill.)
Papyrus *8:* 1463
Paracelsus, Philippus Aureolus *1:* 84
Parasites *8:* **1467-1475,** 1471 (ill.), 1472 (ill.), 1474 (ill.)
Parasitology *8:* 1469
Parathyroid glands *5:* 798
Paré, Ambroise *8:* 1580, *10:* 1855
Parkinson's disease *1:* 62
Parsons, Charles A. *9:* 1820
Particle accelerators *8:* **1475-1482,**

Index

1478 (ill.), 1480 (ill.)
Particulate radiation *8:* 1620
Parturition. *See* **Birth**
Pascal, Blaise *2:* 370, *8:* 1576
Pascaline *2:* 371
Pasteur, Louis *1:* 161, 165, *2:* 292, *10:* 1958, 1990, 1959 (ill.).
Pavlov, Ivan P. *9:* 1657, *10:* 1990
Pelagic zone *7:* 1415
Pellagra *6:* 1219
Penicillin *1:* 155, 156, 157 (ill.)
Peppered Moth *1:* 28
Perception *8:* **1482-1485,** 1483 (ill.)
Periodic function *8:* **1485-1486,** 1485 (ill.)
Periodic table *4:* 777, 778 (ill.), *8:* 1489, **1486-1490,** 1489 (ill.), 1490 (ill.)
Periscopes *10:* 1835
Perrier, C. *4:* 775, *10:* 1913
Perseid meteors *7:* 1263
Persian Gulf War *3:* 462, *7:* 1425
Pesticides *1:* 67, 67 (ill.), 68 (ill.), *4:* 619-622
PET scans *2:* 304, *8:* 1640
Petroglyphs and pictographs *8:* **1491-1492,** 1491 (ill.)
Petroleum *7:* 1418, 1423, *8:* **1492-1495**
Peyote *6:* 1029
Pfleumer, Fritz *6:* 1211
pH *8:* **1495-1497**
Phaeophyta *1:* 95
Phages *10:* 1974
Phenothiazine *10:* 1906
Phenylketonuria *7:* 1254
Phloem *6:* 1175, *8:* 1523
Phobias *8:* **1497-1498**
Phobos *6:* 1229
Phosphates *6:* 1095
Phosphorescence *6:* 1197
Phosphorus *7:* 1347
Photochemistry *8:* **1498-1499**
Photocopying *8:* **1499-1502,** 1500 (ill.), 1501 (ill.)
Photoelectric cell *8:* 1504
Photoelectric effect *6:* 1188, *8:* **1502-1505**
Photoelectric theory *8:* 1503
Photoreceptors *8:* 1484
Photosphere *10:* 1846

Photosynthesis *2:* 306, 388, 391, *8:* **1505-1507**
Phototropism *6:* 1051, *8:* **1508-1510,** 1508 (ill.)
Physical therapy *8:* **1511-1513,** 1511 (ill.)
Physics *8:* **1513-1516**
Physiology *8:* **1516-1518**
Phytoplankton *8:* 1520, 1521
Pia mater *2:* 342
Piazzi, Giuseppe *1:* 201
Pico de Orizaba *7:* 1359
Pictographs, petroglyphs and *8:* 1491-1492
Pigments, dyes and *4:* 686-690
Pineal gland *5:* 798
PKU (phenylketonuria) *7:* 1254
Place value *7:* 1397
Plages *10:* 1846
Plague *8:* **1518-1520,** 1519 (ill.)
Planck's constant *8:* 1504
Plane *6:* 1036 (ill.), 1207
Planetary motion, laws of *7:* 1426
Plankton *8:* **1520-1522**
Plant behavior *2:* 270
Plant hormones *6:* 1051
Plants *1:* 91, *2:* 337, 388, 392, *8:* 1505, **1522-1527,** 1524 (ill.), 1526 (ill.)
Plasma *2:* 326, *7:* 1246
Plasma membrane *3:* 432
Plastic surgery *8:* **1527-1531,** 1530 (ill.), *10:* 1857
Plastics *8:* **1532-1534,** 1532 (ill.)
Plastids *3:* 436
Plate tectonics *8:* **1534-1539,** 1536 (ill.), 1538 (ill.)
Platelets *2:* 329
Platinum *8:* 1566, 1569-1570
Pluto (planet) *8:* **1539-1542,** 1541 (ill.)
Pneumonia *9:* 1681
Pneumonic plague *8:* 1520
Poisons and toxins *8:* **1542-1546,** 1545 (ill.)
Polar and nonpolar bonds *3:* 456
Poliomyelitis *8:* **1546-1549,** 1548 (ill.), *10:* 1958
Pollination *5:* 880-882
Pollution *1:* 9, *8:* **1549-1558,** 1554 (ill.), 1557 (ill.)

Pollution control *8:* **1558-1562,** 1559 (ill.), 1560 (ill.)
Polonium *7:* 1449, 1450
Polygons *8:* **1562-1563,** 1562 (ill.)
Polymers *8:* **1563-1566,** 1565 (ill.)
Polysaccharides *2:* 388
Pompeii *5:* 828, *10:* 1997
Pons, Stanley *7:* 1371
Pontiac Fever *6:* 1181
Pope Gregory XIII *2:* 373
Porpoises *3:* 448
Positive reinforcement. *See* **Reinforcement, positive and negative**
Positron *1:* 163, *4:* 772
Positron-emission tomography. *See* **PET scans**
Positrons *10:* 1832, 1834
Post-it™ notes *1:* 38
Post-traumatic stress disorder *9:* 1826
Potassium *1:* 102
Potassium salts *6:* 1095
Potential difference *4:* 738, 744
Pottery *3:* 447
Praseodymium *6:* 1163
Precambrian era *5:* 988, *8:* 1459
Precious metals *8:* **1566-1570,** 1568 (ill.)
Pregnancy, effect of alcohol on *1:* 87
Pregnancy, Rh factor in *9:* 1684
Pressure *8:* **1570-1571**
Priestley, Joseph *2:* 394, 404, *3:* 525, *7:* 1345, 1444
Primary succession *10:* 1837, 2026
Primates *8:* **1571-1575,** 1573 (ill.), 1574 (ill.)
Probability theory *8:* **1575-1578**
Procaryotae *2:* 253
Progesterone *5:* 800
Projectile motion. *See* **Ballistics**
Prokaryotes *3:* 429
Promethium *6:* 1163
Proof (mathematics) *8:* **1578-1579**
Propanol *1:* 91
Prosthetics *8:* **1579-1583,** 1581 (ill.), 1582 (ill.)
Protease inhibitors *8:* **1583-1586,** 1585 (ill.)
Proteins *7:* 1399, *8:* **1586-1589,** 1588 (ill.)
Protons *10:* 1830, 1832
Protozoa *8:* 1470, **1590-1592,** 1590 (ill.)

Psilocybin *6:* 1029
Psychiatry *8:* **1592-1594**
Psychoanalysis *8:* 1594
Psychology *8:* **1594-1596**
Psychosis *8:* **1596-1598**
Ptolemaic system *3:* 574
Puberty *8:* **1599-1601,** *9:* 1670
Pulley *6:* 1207
Pulsars *7:* 1340
Pyrenees *5:* 826
Pyroclastic flow *10:* 1996
Pyrrophyta *1:* 94
Pythagoras of Samos *8:* 1601
Pythagorean theorem *8:* **1601**
Pytheas *10:* 1890

Q

Qualitative analysis *8:* **1603-1604**
Quantitative analysis *8:* **1604-1607**
Quantum mechanics *8:* **1607-1609**
Quantum number *4:* 772
Quarks *10:* 1830
Quartz *2:* 400
Quasars *8:* **1609-1613,** 1611 (ill.)

R

Rabies *10:* 1958
Radar *8:* **1613-1615,** 1614 (ill.)
Radial keratotomy *8:* **1615-1618,** 1618 (ill.)
Radiation *6:* 1044, *8:* **1619-1621**
Radiation exposure *8:* **1621-1625,** 1623 (ill.), 1625 (ill.)
Radio *8:* **1626-1628,** 1628 (ill.)
Radio astronomy *8:* **1633-1637,** 1635 (ill.)
Radio waves *4:* 765
Radioactive decay dating *4:* 616
Radioactive fallout *7:* 1385, 1386
Radioactive isotopes *6:* 1142, *7:* 1373
Radioactive tracers *8:* **1629-1630**
Radioactivity *8:* **1630-1633**
Radiocarbon dating *1:* 176
Radiology *8:* **1637-1641,** 1640 (ill.)
Radionuclides *7:* 1372
Radiosonde *2:* 216
Radium *1:* 105

Index

Radon   *7:* 1349, 1350
Rain forests   *2:* 295, *8:* **1641-1645,** 1643 (ill.), 1644 (ill.)
Rainbows   *2:* 222
Rainforests   *8:* 1643 (ill.)
Ramjets   *6:* 1144
Rare earth elements   *6:* 1164
Rat-kangaroos   *6:* 1157
Rational numbers   *1:* 180
Rawinsonde   *2:* 216
Reaction, chemical
Reaction, chemical   *9:* **1647-1649,** 1649 (ill.)
Reality engine   *10:* 1969, 1970
Reber, Grote   *8:* 1635
Receptor cells   *8:* 1484
Recommended Dietary Allowances   *10:* 1984
Reconstructive surgery.   *See* **Plastic surgery**
Recycling   *9:* **1650-1653,** 1650 (ill.), 1651 (ill.), *10:* 2009
Red algae   *1:* 94
Red blood cells   *2:* 327, 328 (ill.)
Red giants   *9:* **1653-1654**
Red tides   *1:* 96
Redox reactions   *7:* 1439, 1441
Redshift   *8:* 1611, *9:* **1654-1656,** 1656 (ill.)
Reflector telescopes   *10:* 1871
Refractor telescopes   *10:* 1870
Reines, Frederick   *10:* 1833
Reinforcement, positive and negative   *9:* **1657-1659**
Reis, Johann Philipp   *10:* 1867
Reitz, Bruce   *10:* 1926
Relative dating   *4:* 616
Relative motion   *9:* 1660
Relativity, theory of   *9:* **1659-1664**
Relaxation techniques   *1:* 118
REM sleep   *9:* 1747
Reproduction   *9:* **1664-1667,** 1664 (ill.), 1666 (ill.)
Reproductive system   *9:* **1667-1670,** 1669 (ill.)
Reptiles   *9:* **1670-1672,** 1671 (ill.)
Reptiles, age of   *8:* 1462
Respiration
Respiration   *2:* 392, *9:* **1672-1677**
Respiratory system   *9:* **1677-1683,** 1679 (ill.), 1682 (ill.)

Retroviruses   *10:* 1978
Reye's syndrome   *1:* 8
Rh factor   *9:* **1683-1685,** 1684 (ill.)
Rheumatoid arthritis   *1:* 183
Rhinoplasty   *8:* 1527
Rhodophyta   *1:* 94
Ribonucleic acid   *7:* 1390, 1392 (ill.)
Rickets   *6:* 1219, *7:* 1403
Riemann, Georg Friedrich Bernhard   *10:* 1899
Rift valleys   *7:* 1303
Ritalin   *2:* 238
Rivers   *9:* **1685-1690,** 1687 (ill.), 1689 (ill.)
RNA   *7:* 1390, 1392 (ill.)
Robert Fulton   *10:* 1835
Robotics   *1:* 189, *9:* **1690-1692,** 1691 (ill.)
Robson, Mount   *7:* 1357
Rock carvings and paintings   *8:* 1491
Rock cycle   *9:* 1705
Rockets and missiles   *9:* **1693-1701,** 1695 (ill.), 1697 (ill.), 1780 (ill.)
Rocks   *9:* **1701-1706,** 1701 (ill.), 1703 (ill.), 1704 (ill.)
Rocky Mountains   *7:* 1301, 1357
Roentgen, William   *10:* 2033
Rogers, Carl   *8:* 1596
Root, Elijah King   *7:* 1237
Ross Ice Shelf   *1:* 149
Roundworms   *8:* 1471
RR Lyrae stars   *10:* 1964
RU-486   *3:* 565
Rubidium   *1:* 102
Rural techno-ecosystems   *2:* 302
Rush, Benjamin   *9:* 1713
Rust   *7:* 1442
Rutherford, Daniel   *7:* 1345
Rutherford, Ernest   *2:* 233, *7:* 1337

S

Sabin vaccine   *8:* 1548
Sabin, Albert   *8:* 1549
Sahara Desert   *1:* 52
St. Helens, Mount   *10:* 1996
Salicylic acid   *1:* 6
Salk vaccine   *8:* 1548
Salk, Jonas   *8:* 1548, *10:* 1959
Salyut 1   *9:* 1781, 1788

Samarium *6:* 1163
San Andreas Fault *5:* 854
Sandage, Allan *8:* 1610
Sarcodina *8:* 1592
Satellite television *10:* 1877
Satellites *9:* **1707-1708,** 1707 (ill.)
Saturn (planet) *9:* **1708-1712,** 1709 (ill.), 1710 (ill.)
Savanna *2:* 296
Savants *9:* **1712-1715**
Saxitoxin *2:* 288
Scanning Tunneling Microscopy *10:* 1939
Scaphopoda *7:* 1289
Scheele, Carl *7:* 1345, 1444
Scheele, Karl Wilhelm *3:* 525, *6:* 1032
Schiaparelli, Giovanni *7:* 1263
Schizophrenia *8:* 1596, *9:* **1716-1722,** 1718 (ill.), 1721 (ill.)
Schmidt, Maarten *8:* 1611
Scientific method *9:* **1722-1726**
Scorpions *1:* 169
Screw *6:* 1208
Scurvy *6:* 1218, *10:* 1981, 1989
Seamounts *10:* 1994
Seashore biome *2:* 301
Seasons *9:* **1726-1729,** 1726 (ill.)
Second law of motion *6:* 1171, *7:* 1235
Second law of planetary motion *7:* 1426
Second law of thermodynamics *10:* 1886
Secondary cells *2:* 270
Secondary succession *10:* 1837, *10:* 1838
The Secret of Nature Revealed *5:* 877
Sedimentary rocks *9:* 1703
Seeds *9:* **1729-1733,** 1732 (ill.)
Segré, Emilio *1:* 163, *4:* 775, *6:* 1035, *10:* 1913
Seismic waves *4:* 703
Selenium *7:* 1449
Semaphore *10:* 1864
Semiconductors *4:* 666, 734, *10:* 1910, 1910
Semi-evergreen tropical forest *2:* 298
Senility *4:* 622
Senses and perception *8:* 1482
Septicemia plague *8:* 1519
Serotonin *2:* 350
Serpentines *1:* 191

Sertürner, Friedrich *1:* 33
Set theory *9:* **1733-1735,** 1734 (ill.), 1735 (ill.)
Sexual reproduction *9:* 1666
Sexually transmitted diseases *9:* **1735-1739,** 1737 (ill.), 1738 (ill.)
Shell shock *9:* 1826
Shepard, Alan *9:* 1779
Shockley, William *10:* 1910
Shoemaker, Carolyn *6:* 1151
Shoemaker-Levy 9 (comet) *6:* 1151
Shooting stars. *See* **Meteors and meteorites**
Shumway, Norman *10:* 1926
SI system *10:* 1950
Sickle-cell anemia *2:* 320
SIDS. *See* **Sudden infant death syndrome (SIDS)**
Significance of relativity theory *9:* 1663
Silicon *2:* 400, 401
Silicon carbide *1:* 2
Silver *8:* 1566, 1569
Simpson, James Young *1:* 143
Sitter, Willem de *3:* 575
Skeletal muscles *7:* 1310 (ill.), **1311-1313**
Skeletal system *9:* **1739-1743,** 1740 (ill.), 1742 (ill.)
Skin *2:* 362
Skylab *9:* 1781, 1788
Slash-and-burn agriculture *9:* **1743-1744,** 1744 (ill.)
Sleep and sleep disorders *9:* **1745-1749,** 1748 (ill.)
Sleep apnea *9:* 1749, *10:* 1841
Slipher, Vesto Melvin *9:* 1654
Smallpox *10:* 1957
Smell *9:* **1750-1752,** 1750 (ill.)
Smoking *1:* 34, 119, *3:* 476, *9:* 1682
Smoking (food preservation) *5:* 890
Smooth muscles *7:* 1312
Snakes *9:* **1752-1756,** 1754 (ill.)
Soaps and detergents *9:* **1756-1758**
Sobrero, Ascanio *5:* 844
Sodium *1:* 100, 101 (ill.)
Sodium chloride *6:* 1096
Software *3:* 549-554
Soil *9:* **1758-1762.** 1760 (ill.)
Soil conditioners *1:* 67
Solar activity cycle *10:* 1848

Index

Solar cells *8:* 1504, 1505
Solar eclipses *4:* 724
Solar flares *10:* 1846, 1848 (ill.)
Solar power *1:* 115, 115 (ill.)
Solar system *9:* **1762-1767,** 1764 (ill.), 1766 (ill.)
Solstice *9:* 1728
Solution *9:* **1767-1770**
Somatotropic hormone *5:* 797
Sonar *1:* 22, *9:* **1770-1772**
Sørenson, Søren *8:* 1495
Sound. *See* **Acoustics**
South America *9:* **1772-1776,** 1773 (ill.), 1775 (ill.)
South Asia *1:* 197
Southeast Asia *1:* 199
Space *9:* **1776-1777**
Space probes *9:* **1783-1787,** 1785 (ill.), 1786 (ill.)
Space shuttles *9:* 1782 (ill.), 1783
Space station, international *9:* **1788-1792,** 1789 (ill.)
Space stations *9:* 1781
Space, curvature of *3:* 575, *7:* 1428
Space-filling model *7:* 1286, 1286 (ill.)
Space-time continuum *9:* 1777
Spacecraft, manned *9:* **1777-1783,** 1780 (ill.), 1782 (ill.)
Spacecraft, unmanned *9:* 1783
Specific gravity *4:* 625
Specific heat capacity *6:* 1045
Spectrometer *7:* 1239, 1240 (ill.)
Spectroscopes *9:* 1792
Spectroscopy *9:* **1792-1794,** 1792 (ill.)
Spectrum *9:* 1654, **1794-1796**
Speech *9:* **1796-1799**
Speed of light *6:* 1190
Sperm *4:* 785, *5:* 800, *9:* 1667
Spiders *1:* 169
Spina bifida *2:* 321, 321 (ill.)
Split-brain research *2:* 346
Sponges *9:* **1799-1800,** 1800 (ill.)
Sporozoa *8:* 1592
Sprengel, Christian Konrad *5:* 877
Squid *7:* 1289
Staphylococcus *2:* 258, 289
Star clusters *9:* **1808-1810,** 1808 (ill.)
Starburst galaxies *9:* **1806-1808,** 1806 (ill.)
Stars *9:* **1801-1806,** 1803 (ill.), 1804 (ill.)

binary stars *2:* 276-278
brown dwarf *2:* 358-359
magnetic fields *9:* 1820
variable stars *10:* 1963-1964
white dwarf *10:* 2027-2028
Static electricity *4:* 742
Stationary fronts *1:* 82
Statistics *9:* **1810-1817**
Staudinger, Hermann *8:* 1565
STDs. *See* **Sexually transmitted diseases**
Steam engines *9:* **1817-1820,** 1819 (ill.)
Steel industry *6:* 1098
Stellar magnetic fields *9:* **1820-1823,** 1822 (ill.)
Sterilization *3:* 565
Stomach ulcers *4:* 656
Stone, Edward *1:* 6
Stonehenge *1:* 173, 172 (ill.)
Stoney, George Johnstone *4:* 771
Storm surges *9:* **1823-1826,** 1825 (ill.)
Storm tide *9:* 1824
Strassmann, Fritz *7:* 1361
Stratosphere *2:* 213
Streptomycin *1:* 155
Stress *9:* **1826-1828**
Strike lines *5:* 988
Stroke *2:* 350, 351
Strontium *1:* 105
Subatomic particles *10:* **1829-1834,** 1833 (ill.)
Submarine canyons *3:* 562
Submarines *10:* **1834-1836,** 1836 (ill.)
Subtropical evergreen forests *5:* 908
Succession *10:* **1837-1840,** 1839 (ill.)
Sudden infant death syndrome (SIDS) *10:* **1840-1844**
Sulfa drugs *1:* 156
Sulfur *6:* 1096, *7:* 1446
Sulfur cycle *7:* 1448, 1448 (ill.)
Sulfuric acid *7:* 1447
Sun *10:* **1844-1849,** 1847 (ill.), 1848 (ill.)
  stellar magnetic field *9:* 1821
Sun dogs *2:* 224
Sunspots *6:* 1077
Super Collider *8:* 1482
Superclusters *9:* 1809
Superconducting Super Collider *10:* 1852

Superconductors *4:* 734, *10:* **1849-1852,** 1851 (ill.)
Supernova *9:* 1654, *10:* **1852-1854,** 1854 (ill.)
Supersonic flight *1:* 43
Surgery *8:* 1527-1531, *10:* **1855-1858,** 1857 (ill.), 1858 (ill.)
Swamps *10:* 2024
Swan, Joseph Wilson *6:* 1088
Symbolic logic *10:* **1859-1860**
Synchrotron *8:* 1481
Synchrotron radiation *10:* 2037
Synthesis *9:* 1648
Syphilis *9:* 1736, 1738 (ill.)
Système International d'Unités *10:* 1950
Szent-Györyi, Albert *6:* 1219

T

Tagliacozzi, Gasparo *8:* 1528
Tapeworms *8:* 1472
Tarsiers *8:* 1572, 1573 (ill.)
Tasmania *2:* 241
Taste *10:* **1861-1863,** 1861 (ill.), 1862 (ill.)
Taste buds *10:* 1861 (ill.), 1862
Tay-Sachs disease *2:* 320
TCDD *1:* 54, *4:* 668
TCP/IP *6:* 1126
Tears *5:* 852
Technetium *4:* 775, *10:* 1913
Telegraph *10:* **1863-1866**
Telephone *10:* **1866-1869,** 1867 (ill.)
Telescope *10:* **1869-1875,** 1872 (ill.), 1874 (ill.)
Television *5:* 871, *10:* **1875-1879**
Tellurium *7:* 1449, 1450
Temperate grassland *2:* 296
Temperate forests *2:* 295, *5:* 909, *8:* 1644
Temperature *6:* 1044, *10:* **1879-1882**
Terbium *6:* 1163
Terrestrial biomes *2:* 293
Testes *5:* 800, *8:* 1599, *9:* 1667
Testosterone *8:* 1599
Tetanus *2:* 258
Tetracyclines *1:* 158
Tetrahydrocannabinol *6:* 1224
Textile industry *6:* 1097

Thalamus *2:* 342
Thallium *1:* 126
THC *6:* 1224
Therapy, physical *8:* 1511-1513
Thermal energy *6:* 1044
Thermal expansion *5:* 842-843, *10:* **1883-1884,** 1883 (ill.)
Thermodynamics *10:* **1885-1887**
Thermoluminescence *4:* 618
Thermometers *10:* 1881
Thermonuclear reactions *7:* 1368
Thermoplastic *8:* 1533
Thermosetting plastics *8:* 1533
Thermosphere *2:* 213
Thiamine. *See* **Vitamin B1**
Third law of motion *6:* 1171
Third law of planetary motion *7:* 1426
Thomson, Benjamin *10:* 1885
Thomson, J. J. *2:* 233, *4:* 771
Thomson, William *10:* 1885, 1882
Thorium *1:* 26
Thulium *6:* 1163
Thunder *10:* 1889
Thunderstorms *10:* **1887-1890,** 1889 (ill.)
Thymus *2:* 329, *5:* 798
Thyroxine *6:* 1035
Ticks *1:* 170, *8:* 1475
Tidal and ocean thermal energy *1:* 117
Tides *1:* 117, *10:* **1890-1894,** 1892 (ill.), 1893 (ill.)
Tigers *5:* 859
Time *10:* **1894-1897,** 1896 (ill.)
Tin *2:* 401, 402
TIROS 1 *2:* 217
Titan *9:* 1711
Titania *10:* 1954
Titanic *6:* 1081
Titius, Johann *1:* 201
Tools, hand *6:* 1036
Topology *10:* **1897-1899,** 1898 (ill.), 1899 (ill.)
Tornadoes *10:* **1900-1903,** 1900 (ill.)
Torricelli, Evangelista *2:* 265
Touch *10:* **1903-1905**
Toxins, poisons and *8:* 1542-1546
Tranquilizers *10:* **1905-1908,** 1907 (ill.)
Transformers *10:* **1908-1910,** 1909 (ill.)
Transistors *10:* 1962, **1910-1913,** 1912 (ill.)

Index

U·X·L Encyclopedia of Science, 2nd Edition   **l x i**

Index

Transition elements   *10:* **1913-1923,** 1917 (ill.), 1920 (ill.), 1922 (ill.)
Transplants, surgical   *10:* **1923-1927,** 1926 (ill.)
Transuranium elements   *1:* 24
Transverse wave   *10:* 2015
Tree-ring dating   *4:* 619
Trees   *10:* **1927-1931,** 1928 (ill.)
Trematodes   *8:* 1473
Trenches, ocean   *7:* 1410
Trevithick, Richard   *6:* 1099
Trichomoniasis   *9:* 1735
Trigonometric functions   *10:* 1931
Trigonometry   *10:* **1931-1933**
Triode   *10:* 1961
Triton   *7:* 1332
Tropical evergreen forests   *5:* 908
Tropical grasslands   *2:* 296
Tropical rain forests   *5:* 908, *8:* 1642
Tropism   *2:* 271
Troposphere   *2:* 212
Trusses   *2:* 356
Ts'ai Lun   *8:* 1463
Tularemia   *2:* 289
Tumors   *10:* **1934-1937,** 1934 (ill.), 1936 (ill.)
Tundra   *2:* 293
Tunneling   *10:* **1937-1939,** 1937 (ill.)
Turbojets   *6:* 1146
Turboprop engines   *6:* 1146
Turbulent flow   *1:* 40

U

U.S.S. *Nautilus* 10: *1836*
Ulcers (stomach)   *4:* 656
Ultrasonics   *1:* 23, *10:* **1941-1943,** 1942 (ill.)
Ultrasound   *8:* 1640
Ultraviolet astronomy   *10:* **1943-1946,** 1945 (ill.)
Ultraviolet radiation   *4:* 765
Ultraviolet telescopes   *10:* 1945
Uluru   *2:* 240
Umbriel   *10:* 1954
Uncertainty principle   *8:* 1609
Uniformitarianism   *10:* **1946-1947**
Units and standards   *7:* 1265, *10:* **1948-1952**
Universe, creation of   *2:* 273

Uranium   *1:* 25, *7:* 1361, 1363
Uranus (planet)   *10:* **1952-1955,** 1953 (ill.), 1954 (ill.)
Urban-Industrial techno-ecosystems   *2:* 302
Urea   *4:* 645
Urethra   *5:* 841
Urine   *1:* 139, *5:* 840
Urodeles   *1:* 137
Ussher, James   *10:* 1946

V

Vaccination. *See* **Immunization**
Vaccines   *10:* **1957-1960,** 1959 (ill.)
Vacuoles   *3:* 436
Vacuum   *10:* **1960-1961**
Vacuum tube diode   *4:* 666
Vacuum tubes   *3:* 416, *10:* **1961-1963**
Vail, Alfred   *10:* 1865
Van de Graaff   *4:* 742 (ill.), *8:* 1475
Van de Graaff, Robert Jemison   *8:* 1475
Van Helmont, Jan Baptista   *2:* 337, 393, 404
Variable stars   *10:* **1963-1964**
Vasectomy   *3:* 565
Venereal disease   *9:* 1735
Venter, J. Craig   *6:* 1063
Venus (planet)   *10:* **1964-1967,** 1965 (ill.), 1966 (ill.)
Vertebrates   *10:* **1967-1968,** 1967 (ill.)
Vesalius, Andreas   *1:* 139
Vesicles   *3:* 433
Vibrations, infrasonic   *1:* 18
Video disk recording   *10:* 1969
Video recording   *10:* **1968-1969**
Vidie, Lucien   *2:* 266
Viè, Françoise   *1:* 97
Vietnam War   *1:* 55, *3:* 460
Virtual reality   *10:* **1969-1974,** 1973 (ill.)
Viruses   *10:* **1974-1981,** 1976 (ill.), 1979 (ill.)
Visible spectrum   *2:* 221
Visualization   *1:* 119
Vitamin A   *6:* 1220, *10:* 1984
Vitamin B   *10:* 1986
Vitamin B$_1$ *6:* 1219
Vitamin B$_3$ *6:* 1219
Vitamin C   *6:* 1219, *10:* 1981, 1987, 1988 (ill.)

Vitamin D   *6:* 1219, *10:* 1985
Vitamin E   *10:* 1985
Vitamin K   *10:* 1986
Vitamins   *7:* 1401, *10:* **1981-1989,** 1988 (ill.)
Vitreous humor   *5:* 851
Viviparous animals   *2:* 317
Vivisection   *10:* **1989-1992**
Volcanoes   *7:* 1411, *10:* **1992-1999,** 1997 (ill.), 1998 (ill.)
Volta, Alessandro   *4:* 752, *10:* 1865
Voltaic cells   *3:* 437
Volume   *10:* **1999-2002**
Von Graefe, Karl Ferdinand   *8:* 1527
Vostok   *9:* 1778
Voyager 2   *10:* 1953
Vrba, Elisabeth   *1:* 32

W

Waksman, Selman   *1:* 157
Wallabies, kangaroos and   *6:* 1153-1157
Wallace, Alfred Russell   *5:* 834
War, Peter   *10:* 1924
Warfare, biological.   *2:* 287-290
Warm fronts   *1:* 82
Waste management   *7:* 1379, *10:* **2003-2010,** 2005 (ill.), 2006 (ill.), 2008 (ill.)
Water   *10:* **2010-2014,** 2013 (ill.)
Water cycle.   *See* **Hydrologic cycle**
Water pollution   *8:* 1556, 1561
Watson, James   *3:* 473, *4:* 786, *5:* 973, 980 (ill.), 982, *7:* 1389
Watson, John B.   *8:* 1595
Watt   *4:* 746
Watt, James   *3:* 606, *9:* 1818
Wave motion   *10:* **2014-2017**
Wave theory of light   *6:* 1187
Wavelength   *4:* 763
Waxes   *6:* 1191
Weather   *3:* 608-610, *10:* 1887-1890, 1900-1903, **2017-2020,** 2017 (ill.)
Weather balloons   *2:* 216 (ill.)
Weather forecasting   *10:* **2020-2023,** 2021 (ill.), 2023 (ill.)
Weather, effect of El Niño on   *4:* 782
Wedge   *6:* 1207
Wegener, Alfred   *8:* 1534
Weights and measures.   *See* **Units and standards**
Welding   *4:* 736
Well, Percival   *8:* 1539
Wells, Horace   *1:* 142
Went, Frits   *6:* 1051
Wertheimer, Max   *8:* 1595
Wetlands   *2:* 299
Wetlands   *10:* **2024-2027,** 2024 (ill.)
Whales   *3:* 448
Wheatstone, Charles   *10:* 1865
Wheel   *6:* 1207
White blood cells   *2:* 328, 1085 (ill.)
White dwarf   *10:* **2027-2028,** 2027 (ill.)
Whitney, Eli   *6:* 1098, *7:* 1237
Whole numbers   *1:* 180
Wiles, Andrew J.   *7:* 1394
Willis, Thomas   *4:* 640, *9:* 1718
Wilmut, Ian   *3:* 487
Wilson, Robert   *8:* 1637
Wind   *10:* **2028-2031,** 2030 (ill.)
Wind cells   *2:* 218
Wind power   *1:* 114, 114 (ill.)
Wind shear   *10:* 2031
Withdrawal   *1:* 35
Wöhler, Friedrich   *7:* 1428
Wöhler, Hans   *1:* 124
Wolves   *2:* 383, 383 (ill.)
World Wide Web   *6:* 1128
WORMs   *3:* 533
Wright, Orville   *1:* 75, 77
Wright, Wilbur   *1:* 77
Wundt, Wilhelm   *8:* 1594

X

X rays   *4:* 764, *8:* 1639, *10:* 1855, **2033-2038,** 2035 (ill.), 2036 (ill.)
X-ray astronomy   *10:* **2038-2041,** 2040 (ill.)
X-ray diffraction   *4:* 650
Xanthophyta   *1:* 95
Xenon   *7:* 1349, 1352
Xerography   *8:* 1502
Xerophthalmia   *6:* 1220
Xylem   *6:* 1175, *8:* 1523

Y

Yangtze River   *1:* 199
Yeast   *10:* **2043-2045,** 2044 (ill.)
Yellow-green algae   *1:* 95

Index

Yoga *1:* 119
Young, Thomas *6:* 1113
Ytterbium *6:* 1163

Z

Zeeman effect *9:* 1823
Zeeman-Doppler imaging *9:* 1823
Zehnder, L. *6:* 1116
Zeppelin, Ferdinand von *1:* 75
Zero *10:* **2047-2048**
Zoophobia *8:* 1497
Zooplankton *8:* 1521, 1522
Zosimos of Panopolis *1:* 84
Zweig, George *10:* 1829
Zworykin, Vladimir *10:* 1875
Zygote *4:* 787

DOES NOT CIRCULATE